EMPower
extending mathematical power

Keeping Things in Proportion

Reasoning with Ratios

TEACHER BOOK

TERC
Because math and science build futures

Mary Jane Schmitt, Myriam Steinback,
Tricia Donovan, and Martha Merson

TERC 2067 Massachusetts Avenue
Cambridge, Massachusetts 02140

EMPower Authors / Research and Development Team
Donna Curry, Mary Jane Schmitt, Tricia Donovan, Myriam Steinback, and Martha Merson

Contributors and Reviewers
Pam Meader, Connie Rivera, Heidi Schuler, and Sherry Soares

Technical Team
Production and Design Team: Juania Ashley, Valerie Martin, and Sherry Soares
Copyeditor: Jill Pellarin

Evaluation Team
Brett Consulting Group:
Belle Brett
Marilyn Matzko

EMPower™ was developed at TERC in Cambridge, Massachusetts. This material is based upon work supported by the National Science Foundation under award number ESI-9911410 and by the Education Research Collaborative at TERC. Any opinions, findings, and conclusions or recommendations expressed in this publication are those of the authors and do not necessarily reflect the views of the National Science Foundation.

TERC is a nonprofit made up of teams of math and science education and research experts dedicated to innovation and creative problem solving. At the frontier of theory and practice, TERC's work encompasses research, content and curriculum development, technology innovation, professional development, and program evaluation. TERC has a passion for social justice and strives to create level playing fields for all learners, reaching more than three million students every year.

https://www.terc.edu/empower

Printed in the United States of America

1 2 3 4 5 6 7 8 9 10

ISBN 978-1-7373997-6-6

Send all *EMPower* content inquiries to:

Adult Numeracy Center at TERC
empower@terc.edu

Adult Numeracy Center

Advisory Board

Donna Curry, Equipped for the Future, National Center
Solange Farina, Math Exchange Group, New York City
Linda Kime, Mathematics Association of America and University of Massachusetts, Boston
Ron Kindig, Los Angeles Unified School District (retired)
Kenn Pendleton, General Education Development (GED) Testing Service
Ujwala Samant, National Center for the Study of Adult Learning and Literacy
William Thomas, American Mathematics Association of Two-Year Colleges (AMAYTC) and University of Toledo

Consultants and Reviewers

Scott W. Beckett, Jacksonville State University
Sherry Fraser, Interactive Mathematics Program
Arthur G. Fruhling, Yuba Community College
Nancy R. Johnson, Manatee Community College
Steve Monk, University of Washington
Elizabeth Phillips, Connected Mathematics Program
Mark Rudd, Albuquerque TVI
Karen Sitren, Santa Fe Community College
Cornelia Tierney, TERC
Kevin Wheeler, Three Rivers Community College

EMPower Teacher Participants

Arizona

Gila River Indian Community 21st Century Community Learning Center, Gila River
Andrea Parrella

Illinois

Adult Education Center, Rock Valley College, Rockford
Phyllis Flanagan, Melinda Harrison

Comprehensive Community Solutions, Inc., YouthBuild, Rockford
Ryan Boyce

Township High School District 214 Community Education, Arlington Heights
Kathy Conrad

Massachusetts

Action for Boston Community Development, Inc., LearningWorks, Boston
Kelly Qualman

Adult Learning Center at Middlesex Community College, Bedford
Roberta Froelich

Brockton Adult Learning Center, Brockton
Marilyn Moses

Community Learning Center, Cambridge
Nellie Dedmon, Sylvia Lotspeich Greene, Linda Huntington

Dimock Commuity Health Center, Roxbury
Martha Gray

Harvard Bridge to Learning and Literacy, Cambridge
Judy Hikes, Carol Kolenik

Holyoke Adult Learning Opportunities (HALO) Center, Holyoke
Kelly Martin, Glenn Yarnell

Hyde Square Task Force, Jamaica Plain
Nora O'Connor

Jewish Vocational Services, Boston
Susan Arase, Terry Lerner, Jeff Snyder

Lowell Adult Education Program, Lowell
Barbara Goodridge

Mount Wachusett Community College, Gardner
Patricia Vorfeld

Notre Dame Education Center, South Boston
Esther D. Leonelli

Pioneer Valley Adult Education Center, Northampton
Olu Odusina

Project Place, Boston
Sharon Carey, Maia Hendrickson

Quinsigamond Community College Adult Basic Education Program, Worcester
Cathy Coleman

Read/Write/Now, Springfield
Lee E. Boone, Susanne Campagna, Michelle Faith Brown, Lucille Fandel

Somerville Center for Adult Learning Experiences (SCALE), Somerville
Tom Glannon

X-Cel Adult Education, Dorchester
Don Sands

Maine

Bailey Evening School, Bath
Pam Bessey

Gardiner Adult and Community Education Program
Diann Bailey

Windham Adult Education, Windham
Eva Giles

EMPower Teacher Participants *(continued)*

New Jersey

New Brunswick Public Schools—Adult Learning Center, New Brunswick
Jacqueline Arkoe, Phyllis Boulanger, Sue Helfand, Karen C. Pickering

New York

Adult Learning Center, La Guardia Community College, Long Island City
Norma Andrade, Mark Trushkowsky

BEGIN Managed Programs, New York
Charles Brover, Solange Farina, Santiago Perez, Elizabeth Reddin

Borough of Manhattan Community College/CUNY Adult Basic Education Program, Center for Continuing Education and Workforce Development, New York
Elliot Fink, Mark Lance

City University of New York, Continuing Education, New York
Lisa Simon

Lehman Adult Learning Center, Bronx
Deidre Freeman

SafeSpace-LifeSkills, New York
Jonna Rao

The Adult Learning Center, College of Staten Island, Staten Island
Karen Johnsen

Pennsylvania

Lancaster-Lebanon Intermediate Unit 13/Career Link, Lancaster
Margaret Giordano, Barbara Tyndall

Rhode Island

Dorcas Place, Parent Literacy Center, Providence
Shannon Dolan, Jerelyn Thomas

Tennessee

Hawkins County Adult Education Program, Rogersville
Lisa Mullins

Acknowledgments

Many friends and family members supported the teachers' and *EMPower's* efforts. Thank you, Denise Deagan, Judith Diamond, Cara Dimattia, Ellen McDevitt, Sandy Strunk, and the board of the Adult Numeracy Network. Roberta Froelich, Brad Hamilton, Michael Hanish, David Hayes, Judy Hikes, Alisa Izumi, Esther D. Leonelli, Myra Love, Lambrina Mileva, Luz Rivas, Johanna Schmitt, Rachael Stark, Jonathan Steinback, and Sean Sutherland all made unique and timely contributions to the project.

We appreciate the encouragement and advice from John (Spud) Bradley, *EMPower's* program officer, and Gerhard Salinger of the National Science Foundation.

We are indebted to every adult student who participated in the piloting of *EMPower*. Their honest feedback and suggestions for what worked and what did not work were invaluable. We could not have completed the curriculum without them or their teachers.

Contents

Introduction to the EMPower Series

Background

*Extending Mathematical Power (*EMPower*)* integrates recent mathematics education reform into the field of education for adults and out-of-school youth. *EMPower* was designed especially for those students who return for a second chance at education by enrolling in remedial and adult basic education programs, high school equivalency programs, and developmental programs at community colleges. However, the series is appropriate for a variety of other settings as well, such as high schools, workplaces, and parent and paraprofessional education programs. *EMPower* builds interest and competency in mathematical problem solving and communication.

Over the course of four years (2000–2004), a collaboration of teachers and researchers with expertise in adult numeracy education and K–12 mathematics reform developed and piloted eight contextualized units. These units are organized around four central topics: number and operation sense; patterns, functions, and relations; geometry and measurement; and data and graphs. The *EMPower* program serves as a model for a cohesive mathematics series that offers content consistent with the *Principles and Standards for School Mathematics* (NCTM, 2000), as well as frameworks that are adult-focused, such as the *Equipped for the Future Content Standards* (Stein, 2000), the *Massachusetts ABE Curriculum Frameworks for Mathematics and Numeracy* (Massachusetts Department of Education, 2001), and the *Adult Numeracy Network's Framework for Adult Numeracy Standards* (Curry, Schmitt, & Waldron, 1995). The program fosters a pedagogy of learning for understanding; it embeds teacher support and is transformative, yet realistic, for multilevel classrooms.

EMPower challenges students and teachers to consistently extend their ideas of what it means to do math. The series focuses on mathematical reasoning, communication, and problem solving with a variety of approaches and strategies, not just rote memorization and symbol manipulation. The program fosters a learning community in which students are encouraged to expand their understanding of mathematics through open-ended investigations, working collaboratively, sharing ideas, and discovering multiple ways for solving problems. The goal of *EMPower* is to help people build experience managing the mathematical demands of various life situations, such as finances and commerce, interpretation of news stories, and leisure activities, and to connect those experiences to mathematical principles.

A Focus on Mathematical Content

The *EMPower* series supports students' and teachers' growth by directing attention to significant mathematical understandings.

EMPower emphasizes:

- Data analysis, geometry and measurement, algebra, and number and operations at all student levels.
- Reliance on benchmark numbers—such as powers and multiples of 10, common fractions, and their decimal and percent equivalents—for making mental calculations.
- Early use of calculators to support computation.
- Development of reasoning on proportion and parts of quantities before consideration of formal operations with rational numbers.
- Making decisions about data where students generate, as well as interpret, graphical representations.
- Geometry and measurement based on opportunities to see and touch in developing an understanding of spatial relationships and formulas.
- Leading with patterns and relationships in contextual situations and the representations of these situations with diagrams, tables, graphs, verbal rules, and symbolic notation to develop algebraic competence.

A Focus on Pedagogy

Mathematics is meaningful within a social context. While mathematical truths are universal, the meaning and relevance of numbers changes according to the setting and culture. Therefore, the *EMPower* pedagogy is focused on sets of connected activities that require communication and discourse.

EMPower asks students to

- Work collaboratively with others on open-ended investigations;
- Share strategies orally and in writing; and
- Justify answers in multiple ways.

Key features of activities provide

- Clear mathematical goals
- Contexts that are engaging and useful for young people and adults
- Opportunities to strengthen mathematical language and communication skills
- Various ways of entering and solving problems
- Puzzles that draw students into problems and motivate them to seek a solution.

Overview of EMPower Units

Features of the Teacher Book

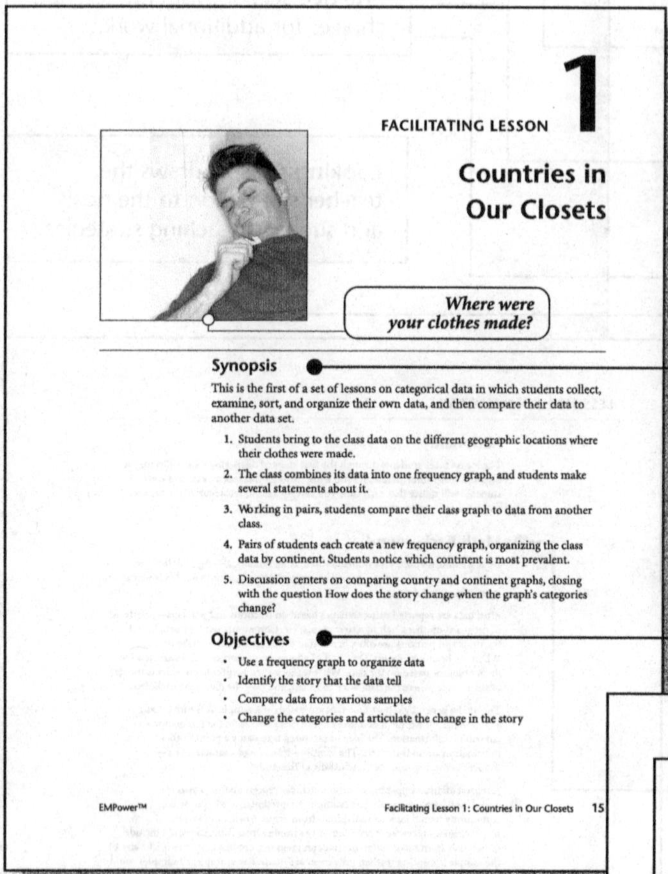

FACILITATING LESSON 1

Countries in Our Closets

Where were your clothes made?

Synopsis

This is the first of a set of lessons on categorical data in which students collect, examine, sort, and organize their own data, and then compare their data to another data set.

1. Students bring to the class data on the different geographic locations where their clothes were made.
2. The class combines its data into one frequency graph, and students make several statements about it.
3. Working in pairs, students compare their class graph to data from another class.
4. Pairs of students each create a new frequency graph, organizing the class data by continent. Students notice which continent is most prevalent.
5. Discussion centers on comparing country and continent graphs, closing with the question How does the story change when the graph's categories change?

Objectives

- Use a frequency graph to organize data
- Identify the story that the data tell
- Compare data from various samples
- Change the categories and articulate the change in the story

EMPower™ Facilitating Lesson 1: Countries in Our Closets 15

Each lesson opens with a **synopsis** of the lesson activities.

Objectives summarize the skills students should gain from the lesson.

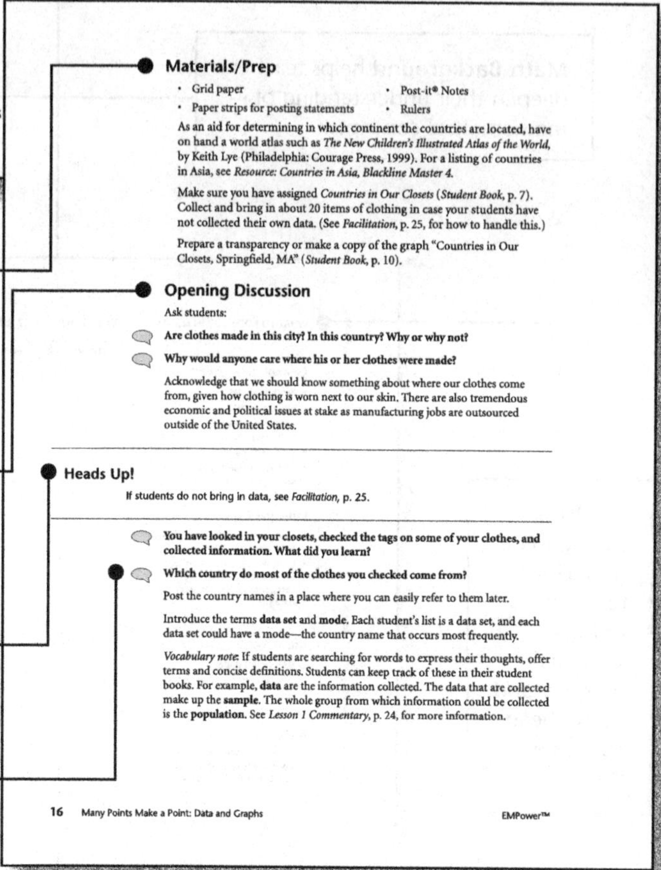

Materials/Prep

- Grid paper
- Paper strips for posting statements
- Post-it® Notes
- Rulers

As an aid for determining in which continent the countries are located, have on hand a world atlas such as *The New Children's Illustrated Atlas of the World*, by Keith Lye (Philadelphia: Courage Press, 1999). For a listing of countries in Asia, see *Resource: Countries in Asia, Blackline Master 4.*

Make sure you have assigned *Countries in Our Closets* (*Student Book*, p. 7). Collect and bring in about 20 items of clothing in case your students have not collected their own data. (See *Facilitation*, p. 25, for how to handle this.)

Prepare a transparency or make a copy of the graph "Countries in Our Closets, Springfield, MA" (*Student Book*, p. 10).

Opening Discussion

Ask students:

Are clothes made in this city? In this country? Why or why not?

Why would anyone care where his or her clothes were made?

Acknowledge that we should know something about where our clothes come from, given how clothing is worn next to our skin. There are also tremendous economic and political issues at stake as manufacturing jobs are outsourced outside of the United States.

Heads Up!

If students do not bring in data, see *Facilitation*, p. 25.

You have looked in your closets, checked the tags on some of your clothes, and collected information. What did you learn?

Which country do most of the clothes you checked come from?

Post the country names in a place where you can easily refer to them later.

Introduce the terms **data set** and **mode**. Each student's list is a data set, and each data set could have a mode—the country name that occurs most frequently.

Vocabulary note: If students are searching for words to express their thoughts, offer terms and concise definitions. Students can keep track of these in their student books. For example, **data** are the information collected. The data that are collected make up the **sample**. The whole group from which information could be collected is the **population**. See *Lesson 1 Commentary*, p. 24, for more information.

16 Many Points Make a Point: Data and Graphs EMPower™

The **Materials/Prep** feature tells teachers what materials they will need to prepare for the day's activities.

The **Opening Discussion** gives teachers ideas for engaging students at the beginning of the lesson.

Heads Up! alerts teachers to helpful hints and things to look for in facilitating the lesson.

Dialog balloons and bold text suggest statements and questions teachers can use to spark and guide class discussion.

A **Summary Discussion** guides the conclusion of the lesson.

Summary Discussion

DATA

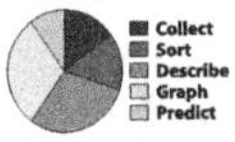

Finish the lesson by posing the following questions. (Students may not have ready answers at this point, but the questions will be a thread through the rest of the lessons.)

What stories can be told from this set of data (or what questions can be asked and answered)?

How does organizing the information into different categories impact how the data may be interpreted?

Who would be curious about what the people in this class eat?

Direct student to the *Reflections* questions (*Student Book*, p. 145).

Practice

Take a Sample, p. 24
For practice choosing a representative sample.

Thirsty and *Ouch!*, pp. 26 and 28
For practice organizing information and making statements.

Extension

Friends and Family Drink Too, p. 30

Data Collection, p. 31
Students begin recording what they eat for a project they will do in a later lesson. For now, they just write down all the foods they eat throughout the day.

Test Practice

Test Practice, p. 32

Looking Closely

Observe whether students are able to

Organize data into consistent categories

Do students group similar foods together? Make sure students are moving index cards around and that they are writing down the category names. If students seem stuck, have them ask you specific questions.

If you see foods categorized incorrectly, ask a question such as "Do all group members agree with the categories?" Encourage students to think, question, and defend their reasoning. If students use overlapping categories such as "cereal" and "grain," present an example close to their own experience. Say, for instance, "It is

EMPower™ Facilitating Lesson 2: Most of Us Eat... 33

Practice, Extension, and **Test Practice** sections list and describe choices for additional work.

Looking Closely draws the teacher's attention to the goals and suggests teaching strategies.

LESSON 1 COMMENTARY

Rationale

The lesson takes students through the first steps of displaying data: collecting and organizing the information. As you work through this lesson, you and your students will notice that each time you categorize and recategorize the data set, you tell a different story.

Math Background

Data are information. This information may be numerical—e.g., salaries, test scores, heights, age, weight—as well as categorical—e.g., countries, foods we eat, types of product defects.

Most data are reported using statistics based on fractions and percents—one-third of pregnant mothers, half of native plants, etc. Commonly used, or benchmark, fractions and percents are often invoked to influence others to make decisions. When the headlines shout, "One-half of all pregnant women ...," readers tend to form opinions based on the data. What readers often neglect to consider is the size of the sample, where the data were collected, and how the data were organized.

The media generally report data that are based on a sample. It is rare for any group to have the time or resources to ask every person in the population to contribute information. The idea of sampling is to study a part in order to gain information about the whole. The sample—its size and characteristics—influences the data and the conclusions of the study.

A representative sample by definition includes representation across the population being surveyed. For example, a representative sample of the community would include individuals from across town, not just one neighborhood. Likewise, a representative sample across America would include individuals from many different states and regions, not just one or two. The size of the sample is important when inferences are made. For example, 30 samples could be sufficient for a population of several hundreds. The deciding factor for determining sample size is how confident you want to be about the inference.

It is not the concern of this unit to determine appropriate sample size, but it is important to note that statisticians use formulas derived from repeated surveys to decide sample size. Regardless, small samples are used in many studies. Factors ... might lead ... outcome.

... k critically ... or into ... ield ... *esson 2.*

EMPower™

Math Background helps teachers deepen their understanding of mathematical concepts.

WHAT TO LOOK FOR IN *LESSON 1*	WHO STANDS OUT? (LIST STUDENTS' INITIALS) STRONG	ADEQUATE	NEEDS WORK	NOTES FOR NEXT STEPS
Concept Development • Records all data • Organizes data • Recognizes key features of frequency graph • Compares data • Changes categories and notes change in story				
Expressive Capacity • Uses precise language to describe the frequency graphs • Contributes to the discussion about the way changing categories affects the story the data tell				
Use of Tools • Uses graph paper • Uses atlas or other reference				
Background Knowledge • Draws on understanding of geography, relative size, and number sense for comparisons				

EMPower™ Facilitating Lesson 1: Countries in Our Closets 23

The **What to Look For** grid allows teachers to track students' progress and think ahead about next steps.

Context

Some students may know about *maquiladoras* in Mexican border towns, where women make clothes for very little money and with no benefits or environmental Occupational Safety and Health Administration (OSHA) workplace protections. CorpWatch (www.corpwatch.org) is one source for information on *maquiladoras*.

Facilitation

If students do not bring in data, or if their sample is too small, skip the second part of the *Opening Discussion*. Have available a pile of 20 clothing articles with labels. First, ask students to predict where the clothes were made. Post the list of their guesses. Note that it will be hard for them to answer this question unless they organize the information on the labels. Then divide up the 20 articles of clothing. Have students write the name of the country for each piece of clothing on a Post-it Note, one country name per note. Ask: "Where are most of our clothes made?" Then continue with the activity.

The authors give ideas for **Making the Lesson Easier** and **Making the Lesson Harder**.

Making the Lesson Easier

Frequency graphs lend themselves to comparisons among categories. If students have little fluency stating comparisons, you may choose only to compare size, using terms like "greater," "fewest," or "less than." For students who are encountering data formally for the first time, the notion that collapsing data yields different stories may be difficult. Treat this lightly in the activity, and revisit such questions after students have more experience categorizing and recategorizing data in the homework and in *Lesson 2*.

Making the Lesson Harder

If your students can handle benchmark fractions and percents, get them to look critically at the data, including the source and sample size. You might ask:

If we asked another class what countries are in their closets, what do you think would happen to the categories? What if we asked the entire community?

How do you think your data would compare to data from another class of adult students in another community?

If students struggle with the idea of sample, you might try this: Have them each write their favorite color on a Post-it Note. If you have a small class, ask them to write the color on two Post-it Notes. Place all of the notes in a container. Have someone randomly (eyes closed) choose a few notes from the container and place them across a line to form a frequency graph. Ask the students how they think this sample compares to the actual total number of colors on notes in the container. You can have them do another frequency graph to compare the sample to the actual total.

Facilitating Lesson 1: Countries in Our Closets **25**

LESSON 1 IN ACTION

Alice articulates the mathematical principle behind compressed data.

I asked, "How did the change in categories affect what we noticed about the data?"

Alice answered, "Well, we keep losing information."

"How so?"

Patiently, Alice explained that when we started our work, every bit of data was visible. She added that we had lost details initially recorded. "At first, we knew every country in every person's closet and how many pieces of clothing came from that country. Then we combined the data, and we lost track of who had which countries. Then we did it by continent, and we lost track of all the countries."

Alice's realization quickly gained agreement from the rest of the class. After all, just the previous week a classmate had noted, "When you change the amount of data you look at, you find different things."

Sonia added her comment with increased conviction: "It is like politics. Politicians use a graph and tell you this is true, but you look at the graph, and it does not tell you everything."

Tricia Donovan
Pioneer Valley Adult Education Center, Northampton, MA

In **Lesson in Action,** EMPower teachers share their classroom experiences.

26 Many Points Make a Point: Data and Graphs EMPower™

Overview of EMPower Units

Features of the Student Book

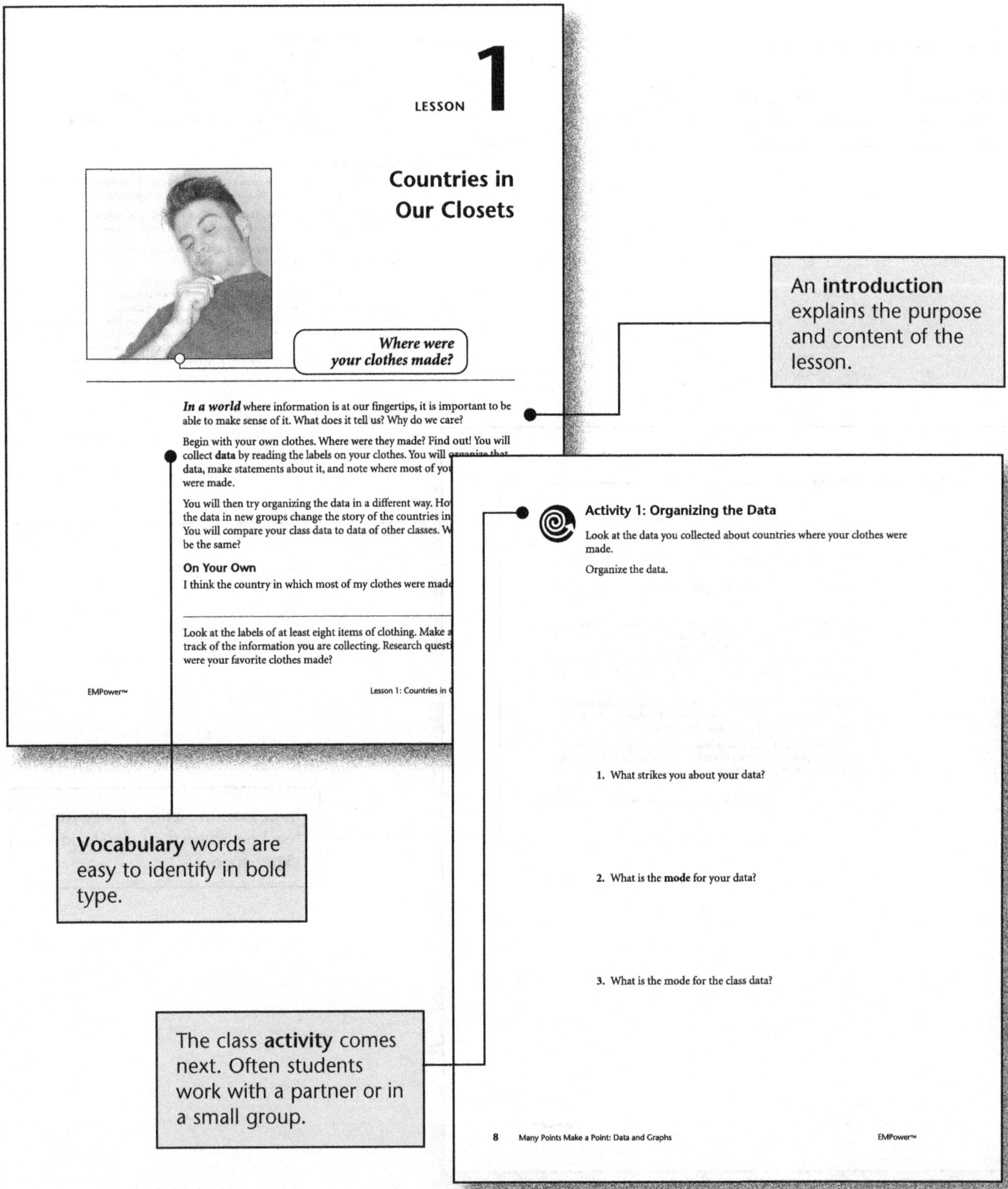

LESSON 1

Countries in Our Closets

Where were your clothes made?

In a world where information is at our fingertips, it is important to be able to make sense of it. What does it tell us? Why do we care?

Begin with your own clothes. Where were they made? Find out! You will collect **data** by reading the labels on your clothes. You will organize that data, make statements about it, and note where most of yo[...] were made.

You will then try organizing the data in a different way. Ho[...] the data in new groups change the story of the countries in[...] You will compare your class data to data of other classes. W[...] be the same?

On Your Own

I think the country in which most of my clothes were mad[...]

Look at the labels of at least eight items of clothing. Make a[...] track of the information you are collecting. Research questi[...] were your favorite clothes made?

EMPower™ Lesson 1: Countries in [...]

Activity 1: Organizing the Data

Look at the data you collected about countries where your clothes were made.

Organize the data.

1. What strikes you about your data?

2. What is the **mode** for your data?

3. What is the mode for the class data?

8 Many Points Make a Point: Data and Graphs EMPower™

Practice: Categorically Speaking

Sometimes we can organize categories into subgroups of other categories in order to better understand the data. For example, drug stores will often organize their aisles by category—skin care, beauty aids, and so on. Then the items in each of those aisles are also organized. This makes it easier to take inventory and place orders and for the customer to find products.

Can you think of other examples where information is categorized?

Use the chart below *or* create your own chart; draw a picture; make a list; or write about examples you see at home or at work. In the chart below, one example is given.

Location	What Stuff or Information?	How Is It Organized?
Drug store	Over-the-counter products they sell	Skin care Stationery Beauty aids First aid Seasonal Hair care

EMPower™ Lesson 1

Practice pages invite students to use new skills.

Extension: Taking Inventory

Find a drawer or closet filled with many things. Use a frequency graph to show the contents. Start with six categories. Then do another frequency graph to show the same items, but this time use three categories.

For example:

My Medicine Cabinet

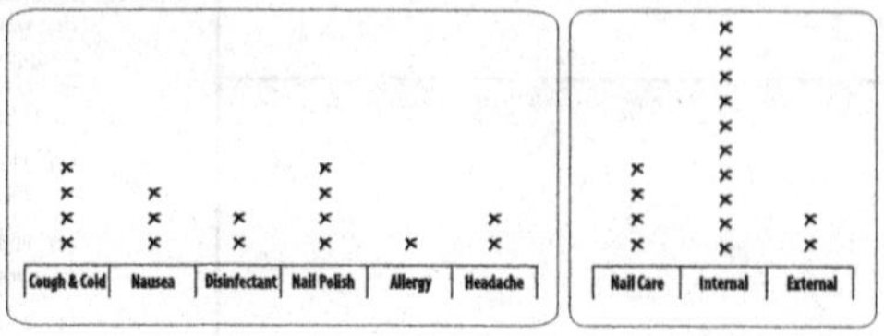

Extension pages present a challenge or a new but related math idea.

Ample space allows students to record most work in their books.

16 Many Points Make a Point: Data and Graphs EMPower™

Test Practice

A **Test Practice** that reflects the format of the GED test concludes each lesson.

1. Freda worked at a printing press. She kept a tally of all the defects she found in the books she was processing. Based on the frequency graph she created, shownbelow, which of the following statements is true?
   (1) There were more defects related to colors than any other defect.
   (2) There were twice as many torn pages as there were warped spines.
   (3) There were more warped spines than colors that bled.
   (4) Half of all the defects were related to colors that bled.
   (5) Half of all the defects were related to pages that were torn.

Book Defects Frequency Graph

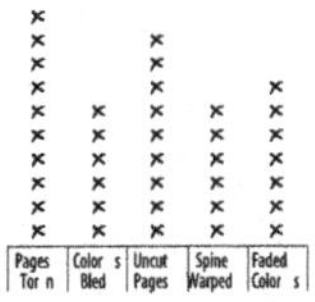

2. Tony, a tour guide, has been keeping a tally of visitors from different states. According to his tally, which of the following is *not* a true statement?
   (1) The fewest number of visitors came from Oregon.
   (2) There were as many visitors from Maine as there were from Georgia.
   (3) There were as many visitors from Georgia as there were from Oregon and Texas.
   (4) There were twice as many visitors from Florida as there were from Rhode Island.
   (5) There were twice as many visitors from Idaho as there were from Maine.

U.S. Tourist Frequency Graph

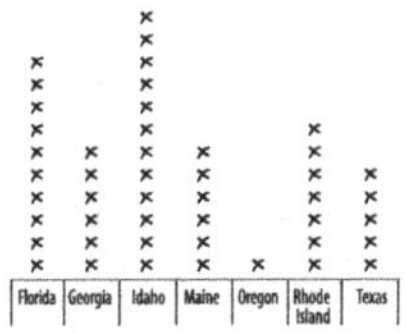

3. Clara is paid a commission on each of the large electronics devices that she sells. She tallied the different items she sold for the month. Based on her tally, what can she tell her boss about her sales?
   (1) She sold twice as many DVD players as she did TVs.
   (2) She sold twice as many DVD players as she did computers.
   (3) She sold more DVD players than she did TVs and computers combined.
   (4) She sold more TVs and VCRs than she did stereos and DVD players combined.
   (5) She sold half as many TVs as she did stereos.

Home Media Frequency Graph

EMPower™

4. Bronson, a forest ranger in the Green Mountains, kept track of the different animals that were reportedly seen in one of the campgrounds during the month of May. Based on his tally, which of the following statements could he tell the media?
   (1) There were twice as many wolves reportedly seen as bears.
   (2) Half of all the reported sightings were wolves.
   (3) There were more bears than wolves reportedly seen.
   (4) There were twice as many moose reportedly seen than there were black bears.
   (5) One-quarter of all reported sightings were brown bears.

North American Wildlife Frequency Graph

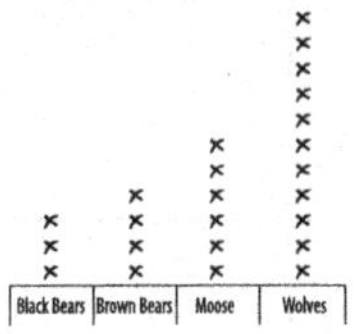

5. According to Global Exchange, some workers in China who make clothing for Disney are paid as little as $0.16 per hour. At this wage, what amount would a worker make for a 40-hour workweek?
   (1) $64.00
   (2) $6.40
   (3) $6.00
   (4) $80.00
   (5) $0.60

6. In 2003 the U.S. Department of Labor (DOL) issued new regulations for overtime pay. The DOL estimated that under the new regulations 1.3 million low-wage workers would become eligible for overtime pay, unless their wages were raised to $425 per week. The DOL estimated that 24.8 percent of those workers were Hispanic and 16.6 percent were African American. What percent were neither Hispanic nor African American?

18 Many Points Make a Point: Data and Graphs EMPower™

Changing the Culture

The authors have created this series to follow the National Council of Teachers of Mathematics (NCTM) Principles and Standards; however, every teacher who uses *EMPower* faces the challenge of transforming the prevailing culture of his or her math classroom. *EMPower* pilot teachers offer some ideas for facilitating this transition:

- Set the stage. As a class, set ground rules. Explicitly state that this is a space for everyone to learn. As one teacher said, "We are in this together. Share, even if you do not think you are right. Whatever you add will be helpful. It lets us see how you are looking at things."
- Group your students. Match students whose learning styles and background knowledge complement each other. Ask questions, such as, "How did it go to work together?" "How did everyone contribute?"
- Allow wait time. Studies have shown that teachers often wait less than three seconds before asking another question. Students need more time to think.
- Sit down. Watch students before interrupting to help them. Listen for logic and evidence of understanding. Follow the thread of students' thinking to uncover unconventional approaches. During discussions with the whole group, hand over the chalk.
- Review written work. Look beyond right and wrong answers to learn everything you can about what a student knows. Determine what seems solid and easy, as well as patterns in errors. If students are scattered, suggest ways they can organize their work; this is likely to lead to more efficient problem solving and clearer communication.
- Question. Hearing the right answer is not necessarily a cue to move on. Question students at this point too. Specific questions are included in the lesson facilitation.

Unit Sequences and Connections

The sequence in which the *EMPower* units can be used effectively with your class will depend on the backgrounds and interests of students. The units are not numbered so teachers can order them according to their class needs; however, the authors suggest specific unit arrangements that will support students' progression through certain concepts.

The authors do not recommend sequencing the units according to the traditional basic math model that begins with whole numbers, follows with fractions, decimals, and percents; data and graphs; algebra; and then geometry. Instead, the authors suggest you integrate the five units that focus on numbers with the units on geometry, data, and algebra. The authors found this integration of topics helped to motivate the adult students in their pilot classes.

Although the units were not specifically designed to build on one another, there are clear connections between some of the units in the series. *Over, Around, and Within: Geometry and Measurement* provides a nice introduction to the program because it focuses on small whole numbers. *Everyday Number Sense: Mental Math and Visual Models* could follow to further develop whole number mental math skills and visual models. *Using Benchmarks: Fractions and Operations* provides the necessary groundwork with fractions, decimals, and percents to describe approximate relationships between data sets in *Many Points Make a Point: Data and Graphs.* And *Split It Up: More Fractions, Decimals, and Percents* continues to expand students' repertoire of familiar fractions, decimals, and percents. *Seeking Patterns, Building Rules: Algebraic Thinking* builds upon the tools and relationships used in *Keeping Things in Proportion: Reasoning with Ratios.*

The following diagram demonstrates this integrated sequence:

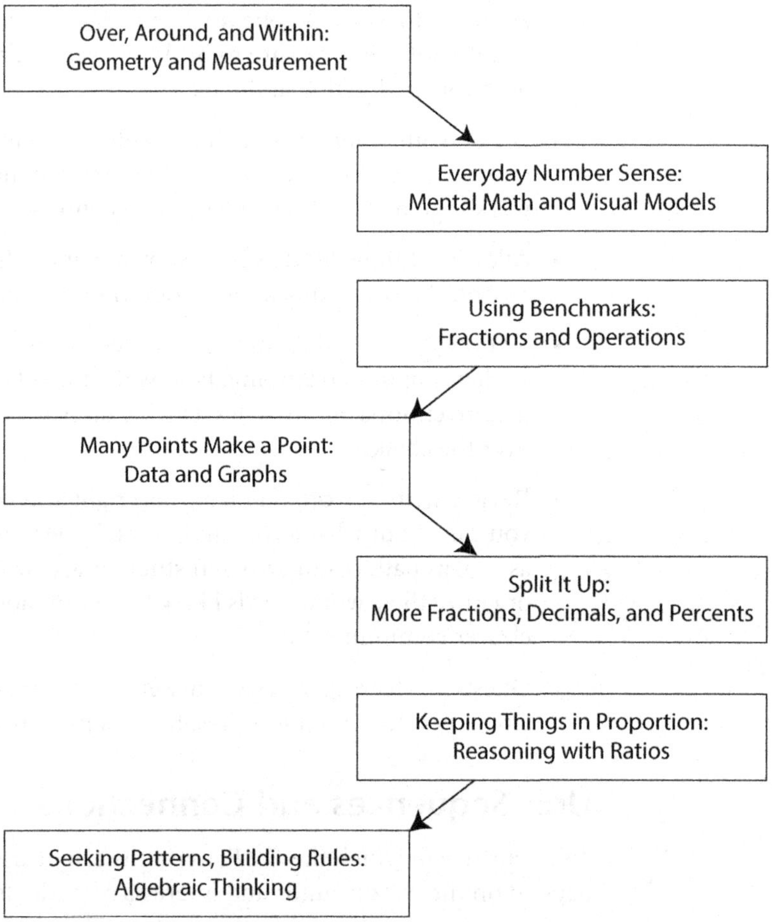

Unit Descriptions

Over, Around, and Within: Geometry and Measurement
Students explore the features and measures of basic shapes. Perimeter and area of two-dimensional shapes and volume of rectangular solids provide the focus.

Everyday Number Sense: Mental Math and Visual Models
Students solve problems and compute with whole numbers using mental math strategies with benchmarks of 1, 10, 100, and 1,000. Number lines, arrays, and diagrams support their conceptual understanding of number relationships and the four operations.

Using Benchmarks: Fractions and Operations
Students use the fractions $\frac{1}{2}$, $\frac{1}{4}$, $\frac{3}{4}$, and $\frac{1}{10}$; the decimals 0.1, 0.5, 0.25, and 0.75; and the percents 50%, 25%, 75%, 100%, and the multiples of 10% as benchmarks to describe and compare all part-whole relationships.

Many Points Make a Point: Data and Graphs
Students collect, organize, and represent data using frequency, bar, and circle graphs. They use line graphs to describe change over time. They use benchmark fractions and the three measures of central tendency—mode, median, and mean—to describe sets of data.

Split It Up: More Fractions, Decimals, and Percents
Building on their command of common benchmark fractions, students add thirds, eighths, hundredths and their decimal and percent equivalents to their repertoire of part-whole relationships.

Keeping Things in Proportion: Reasoning with Ratios
Students use various tools—objects, diagrams, tables, graphs, and equations—to understand proportional and nonproportional relationships.

Seeking Patterns, Building Rules: Algebraic Thinking
Students use a variety of representational tools—diagrams, words, tables, graphs, and equations—to understand linear patterns and functions. They connect the rate of change with the slope of a line and compare linear with nonlinear relationships. They also gain facility with and comprehension of basic algebraic notation.

Frequently Asked Questions

Q: *I have classes that are widely multilevel. Can this work?*

A: Many teachers see a wide range of levels within the group as an obstacle. Turn the range of levels to your advantage. Focus on students' representations (words, graphs, equations, sketches). This gives everyone the chance to see that answers emerge in several ways. Slowing down deepens understanding and avoids facile responses. Having calculators available can even the playing field. Implement the suggestions in *Making the Lesson Easier* and *Making the Lesson Harder* of each lesson facilitation in the *Lesson Commentary* sections.

Q: *How do I deal with erratic attendance patterns?*

A: Uneven attendance can be disruptive. Students who miss class may feel disoriented; however, the lessons spiral back to the most important concepts. When the series circles back, students will have a chance to revisit concepts and get a toehold.

Q: *What do I do if I run out of time, and there is no way to finish a lesson?*

A: Each activity is important, but reviewing it is equally important. It is better to cut the activity short so there is time to talk with students about what they noticed. Maximize the time by selecting a student or group whose work you feel will add to the class's understanding to report their findings. Be conscious of when you are letting an activity go on too long because the energy is high. Fun is good, but be sure important learning is happening. If you like to give time in class to reviewing homework, and you want to hear from everyone in discussions, you will run out of time. Schedule a catch-up session every three or four lessons.

Q: *How do I respond to comments such as "Can't we go back to the old way?"*

A: Change is unsettling, especially for students who are accustomed to math classes where their job is to work silently on a worksheet solving problems by following a straightforward example. Be clear about the reasons why you have chosen to de-emphasize some of the traditional ways of teaching in favor of this approach.

Ultimately, you may need to agree to some changes to accommodate students' input. Meanwhile, stick with the program. Reiterate for students what they have accomplished. When there is an "Aha!" moment, point it out.

Q: *My own math background is not strong. Will I be able to teach this series?*

A: Yes! Most teachers tend to teach the way they were taught. Adopting a different stance requires support, and the more types of support, the better. This series offers support in a few ways. The teacher books for each unit list open-ended questions designed to keep the math on track. In the *Lesson Commentary* sections, *Math Background* helps teachers deepen their understanding of a concept. In addition, the *Lesson in Action* sections provide examples of student work with comments that illuminate the underlying mathematics.

The best support often comes from a colleague. If no one at your site is currently teaching *EMPower*, join the Adult Numeracy Network, http://adultnumeracynetwork.org, and attend your regional NCTM conference. Look for others who are integrating the NCTM Principles and Standards through the use of a curriculum such as *Investigations in Number, Data, and Space* (TERC, 1997); *Connected Mathematics* (Lappan et al., 1998); or *Interactive Mathematics Program* (Alper et al., 1997).

Unit Introduction
Keeping Things in Proportion: Reasoning with Ratios

What is proportional reasoning?

> *Understanding the underlying relationships in a proportional situation and working with these relationships has come to be called proportional reasoning ... Proportional reasoning has been described as the capstone of elementary school arithmetic and the gateway to higher mathematics, including algebra, geometry, and probability, statistics, and certain aspects of discrete mathematics (*National Research Council, 2001, p. 241*).*

Proportional reasoning involves **ratios**. "A ratio is a relational number that has two properties:

1. It relates two quantities in one situation, and
2. It projects that relationship into a second situation in which the relative amounts of the two quantities remain the same" (Smith, 2002, p. 14).

Here's an example: You are shopping and you see a sign proclaiming, "$3 for a dozen ears of corn," but you only want four ears. How could you reason about this? In the first situation, the ratio, or relational number, 12 ears for $3 (12/3), is a guide that allows you to project to a second situation: 4 ears for $1 (4/1).

What is it that enables people to project from that first relationship to the other? What helps people develop an ability to grasp and work with the relationship? How do people develop this kind of mathematical insight, and what kinds of tools are available to support their thinking? These questions and their answers guided the development of the lessons in this book.

Major Themes

The importance of proportional reasoning

A good sense of proportion is required to respond to many mathematical situations adults encounter in daily life. Simple proportion, sometimes called **direct proportion**, is the underlying structure of several everyday adult activities:

- **Buying and selling** often involves knowing the price of a certain quantity and needing to find the price for another larger or smaller amount. *At 8 lb. for $4.00, what would one pay for 13 lb.?* Savvy supermarket shoppers, faced with a dizzying set of choices, make judgments about best buys.
- **Adjusting a recipe** usually requires keeping all amounts in proportion. *A recipe calling for 6 c. flour, 2 c. sugar, and 1½ sticks of butter serves 5 people. What quantities of those ingredients would be needed to make the recipe to serve a dinner party of 20 guests?*

- **Taxes and tips** are proportional to amount spent. *If you pay a 7½ cents sales tax for each $1.00 spent, what is the sales tax for $210?*
- **Graphic design** requires a sense of scale for enlarging or reducing two-dimensional objects. *Would an 8″ by 10″ photo be a proportional enlargement of a 4″ by 6″ photo?*
- **Conversion between measurement units**, including converting between different countries' monetary units, is frequently necessary. *At the current exchange rate, how much would a pair of sneakers cost in Japanese yen?*
- **Probability and risk** are communicated in ratios. *One in eight women will develop breast cancer in her lifetime.*

Although a sense of direct proportion is applicable to many situations, there are many cases where the application of simple proportional reasoning alone does *not* suffice. In this unit, students have opportunities to explore situations when direct proportion is applicable and those when it is not.

Many ways to reason about proportions and many tools to use

Proportional reasoning is a mathematically rich topic. At its heart is multiplicative reasoning. However, there are many ways to think about and solve problems dealing with proportion, and researchers who observe adults going about everyday activities have noticed that people use a variety of strategies. With this in mind, students are encouraged to develop and use a bank of appropriate strategies. Some of these tools include the intuitive and visual; an examination of multiplicative relationships within and between ratios; connection to a corresponding graph; unit rate; and the cross-multiplication property.

In helping students develop their proportional reasoning, it is advisable to delay the cross-multiplication algorithm until other methods that emphasize the underlying meaning of proportional relationship are tried. Smith and Weinberg are two educational researchers who strongly make this point.

> *The central challenge of developing students' capacity to think with ratios (to reason proportionately) is to teach ideas and restrain the quick path to computation. The research literature on ratio and proportional reasoning shows that students' thinking in situations that we see as proportional often shows they are not thinking about "what is going on" in those situations. In their rush to compute an answer, students do not seem to think long enough about how the quantities in the situation relate and change* (Smith, p. 19).

> *As mathematics teachers, we know that it is imperative for students not only to find the answer but also to determine what they did mathematically and why .... Although some textbooks emphasize the cross-multiply strategy for solving proportions, they rarely explain why it should be used .... We must help students realize that many proportional reasoning strategies are appropriate and valid, such as finding a unit-rate, size-change, and equivalent fractions strategies. We must not only teach these strategies to our students, but also reinforce the underlying meaning of the proportional reasoning in the situation. If we help students learn multiple approaches to solving proportional reasoning problems, we accomplish three things: (1) We focus attention on connecting mathematics with the real world; (2) We focus attention on connecting different concepts and skills within mathematics; (3) We reinforce the themes of problem solving, communication, and reasoning by identifying the strengths and weaknesses of different strategies* (Weinberg, 2002, p. 144).

The following graphic communicates the sequence in which the multiple tools are introduced as the lessons progress.

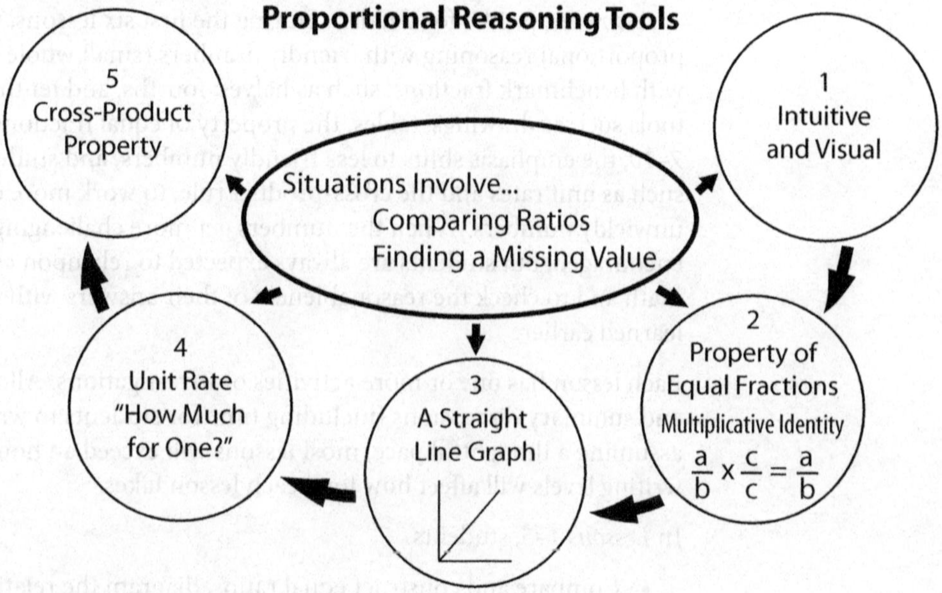

Varying levels of difficulty in proportion situations

Researchers of children's mathematical thinking have found that the relative difficulty of problems depends upon three factors:

1. Familiarity of context,
2. Kind of number (small whole numbers are easier than large whole numbers or fractions and decimals), and
3. The character of the relationship between and within the quantities (double, half, and triple are easy to see) (Lamon, 1993).

The unit as a whole takes these factors into account. *Lesson 1* starts out with a very familiar situation: shopping. The unit slowly builds from comfortable numbers with simple relationships to larger, more difficult numbers with relatively more complicated relationships.

Unit Goals

By the end of this unit, students should be able to

- Use a variety of strategies to examine the relationship within a ratio as well as the relationship between ratios
- Determine whether ratios are equal
- Solve for the missing value when three of four amounts are known
- Determine whether a situation is proportional

The Flow of the Unit

The *Teacher Book* provides materials for initial, interim, and final assessments (see *Assessments*, p. 145, for details). During the first six lessons, the emphasis is on proportional reasoning with friendly numbers (small whole numbers or numbers with benchmark fractions, such as halves, fourths, and tenths). Students use basic tools such as drawings, tables, the property of equal fractions, and graphs. In *Lessons 7–10*, the emphasis shifts to less friendly numbers, and students learn other tools, such as unit rates and the cross-product rule, to work more efficiently with these unwieldy numbers. When the numbers get more challenging, calculator use is encouraged, but students are always expected to rely upon estimation and mental math and to check the reasonableness of their answers with other methods they learned earlier.

Each lesson has one or more activities or investigations. Allowing time for opening and summary discussions (including time for students to write reflections) and assuming a thoughtful pace, most lessons will exceed an hour. Students' reading and writing levels will affect how long each lesson takes.

In *Lessons 1–3*, students

- Compare and construct equal ratios, diagram the relationships, write rules for creating equal ratios, and estimate and predict using ratios involved in shopping, cooking, and figuring the time it takes to do tasks.

In *Lesson 4*, students

- Examine the difference between ratios representing part:part and those representing part:whole in surveys. They focus on the various ways to express ratios.

In *Lessons 5–6*, students

- Work with relationships involved in resizing and redesigning two-dimensional objects (such as replicas of the *Mona Lisa* painting and their personal calculators). They add graphing as a tool for examining the proportions.

In *Lessons 7–9*, students

- Explore the new tools of unit rate and the cross-product property in order to compare and create ratios with more difficult numbers. They work with speed and currency conversions.

In *Lesson 10*, students

- Use percents as a convenient way to compare the relationships within and between data sets of different sizes. National and local demographics of educational programs provide the context.

Assessments

Keeping Things in Proportion: Reasoning with Ratios opens and closes with assessment. In both cases, the *Teacher Book* provides multiple ways to gauge what students know. In addition, the *Teacher Book* includes an *Interim Assessment.* There are also tools for ongoing assessments in each lesson. These components are described next.

Opening the Unit: Comparing and Predicting

What Would It Take?: Students consider what it would take to fulfill their dreams and then share their ideas with a partner. They look at comparisons of each other's desired salaries over time, which requires the use of proportional reasoning.

Initial Assessment: Comparing and Predicting: Students complete several tasks involving proportional reasoning. Checked against the *Initial Assessment Checklist*, the *Initial Assessment* will reveal the strategies students have used to deal with each task.

Rate, Ratio, and Proportion Mind Map: Students record words and ideas they associate with rate, ratio, and proportion. By the end of the unit, their Mind Maps will likely include more examples and terms.

Interim Assessment: Checking In

The *Interim Assessment* provides an opportunity to synthesize learning from the first six lessons. There are four sections.

A Puzzling Situation: Students work in groups of three to resize the pieces of a puzzle.

Recipe Project: Students use a favorite recipe and vary the ingredients for different serving amounts. Then they compare two of the ingredients as they create sets of equal ratios that they use to create a graph.

How Low Can You Go?: Students use flyers from three different stores to compare prices and find the lowest price for each of the items listed.

Test Practice: Students are presented with some multiple choice problems, similar in format to those found on tests such as the GED.

Closing the Unit: Reasoning with Ratios

Rates and Ratios in Your Life: Students chose a context outside of the classroom in which to investigate ratios. They prepare a report to share with the rest of the class.

Final Assessment: As they did in the *Initial Assessment*, students complete tasks and multiple-choice questions, revealing their own assessment of their familiarity with the material. Checked against the *Final Assessment Checklist*, the *Final Assessment* provides information on which skills are solid and which will need further development.

Rate, Ratio, and Proportion Mind Map: Students record words and ideas they associate with rates, ratios, and proportions. Their Mind Maps will likely include more examples and more terms, particularly technical ones, than were in their first Mind Maps in *Opening the Unit*.

Ongoing Assessment

Much of the work students do will take place in small groups or pairs. This work must be assessed to determine whether students are absorbing the main ideas and to diagnose difficulties, as well as to provide more challenging work to higher level students. The *Looking Closely* section of each lesson focuses teachers' attention on objectives and the corresponding observable behavior. Some teachers read this after they have taught the lesson and return to class prepared to assist students who struggled. Teachers may also revisit ideas that were not clearly understood.

As students gain skills, the teacher will want to track their progress and communicate observations to them. Each lesson has a chart called *What to Look For in Lesson #*, with four dimensions for assessing students' progress:

Concept Development—the ability to grasp big ideas and understand multiplicative relationships.

Expressive Capacity—the ability to describe, reason aloud, and explain to others.

Use of Tools—the ability to use calculators, grid paper, manipulatives (such as square-inch tiles), and mathematical methods such as the property of equal fractions, unit rate, or the cross-product property, to communicate and enhance understanding.

Background Knowledge—the ability to draw on and share past experiences using proportional reasoning to make decisions or predictions and to influence or communicate information.

As students work through this unit, they will understand that there are several tools for reasoning proportionally and will learn which tools are most useful in particular situations. They will share knowledge gained in and out of the classroom to become proficient at using a variety of tools to reason with ratios.

Facilitating Opening the Unit: Comparing and Predicting

What is your sense of proportion?

Synopsis

In this session, you will assess students' facility with proportional reasoning, beginning with contexts that may be quite familiar to them. The activities will help establish baseline information about students' strategies, number sense, and background knowledge, and will set the stage for later work with rate, ratio, and proportional and nonproportional situations.

1. Students write about the annual salary they would need to fulfill their dreams, and pairs of students interview one another about how that salary would play out financially. Student pairs prepare a poster, comparing the salaries each would need to fulfill his or her dreams.
2. Students complete a self-assessment based on five tasks. They complete all or some of the problems in the *Initial Assessment.*
3. Together, students brainstorm what the words rate, ratio, and proportion bring to mind. They complete a Mind Map, first individually and then as a class.
4. A discussion of *Unit Goals* and personal goals reinforces the scope of the unit. Tips for success are offered.

Objectives

- Demonstrate ways to compare amounts, either additively (e.g., \$300 is \$225 more than \$75) or multiplicatively (e.g., \$300 is four times as much as \$75, or \$75 is 25% of \$300)
- Demonstrate ability to solve various proportional and nonproportional problems and describe solution strategies
- Share experiences with rate, ratio, and proportion
- Understand the scope of the unit

Materials/Prep

- Calculators
- Markers
- Easel paper
- Sticky notes

Plan to use *What to Look For in Opening the Unit*, p. 7 to record your observations.

Prepare a class vocabulary list to display and add to throughout the unit.

Prepare a shared Mind Map (*Student Book*, p. 4) to display for student viewing and contributions.

Prepare to display the *Initial Assessment*, pp. 145–149, during the assessment. Prepare one *Initial Assessment* for each student to take.

Plan one *Initial Assessment Class Tally*, p. 150, for you to use when you analyze the *Initial Assessment.*

Prepare the *Initial Assessment Checklist*, p. 151, for feedback to each student's individual assessments.

Opening Discussion

Today you want to welcome students to the unit and also gather information about their interpretations of problems and current methods for approaching them.

By the end of *Opening the Unit,* you will have several pieces of evidence of students' proportional reasoning:

- *What Would It Take?* (*Activity 1*)
- *Initial Assessment* (*Activity 2*)
- Mind Map (*Activity* 3)
- The start of a class vocabulary list

Open by saying:

Today at the start of the unit *Keeping Things in Proportion: Reasoning with Ratios,* I do not intend to spend our class time teaching. Rather, you are going to share with each other and me ways that you use to think through a number of mathematical situations that involve comparing and predicting. There is no one way to solve these problems; use whatever method makes the most sense and is most efficient, but be prepared to explain your reasoning.

Activity 1: What Would It Take?

In this first activity, students will start by sharing what they think it would take to fulfill their dreams. Refer students to *Activity 1: What Would It Take?* (*Student Book*, p. 2.)

Say:

Most of us have dreams. What are yours? Think for a few minutes and then write about them. Next consider how large a salary you would need to earn yearly to fulfill those dreams.

When everyone is done writing, ask students to pair up and take turns interviewing one another, using the set of questions provided in *Student Book*, p. 3.

Say:

Listen carefully. When you are the interviewer, take notes for each answer and the way your partner arrived at the answer.

As pairs complete the interviews, say:

When you have each completed the interview, use the information to make a poster with at least three statements that compare your dream salaries. Use fractions, decimals, percents, or differences to show how your dream salaries compare.

As students prepare their comparisons for display, notice the ways they are comparing. Encourage them to use statements that are exact or approximate. For example, "Adeebah needs only half of what Edgardo needs."

Display the completed comparisons. Ask pairs to take a "gallery walk" to examine the comparison information. What is interesting? Is anything confusing? Are the statements accurate? Ask students to leave comments or questions on other pairs' displays.

Pull the class together to read and discuss some of the comments.

Notice in particular how salaries were compared.

Ask:

Who used differences as a way to make comparisons such as "Edgardo's dream salary of $100,000 is $55,000 per year more than Adeebah's dream salary of $45,000" or "Adeebah needs $55,000 less than Edgardo to live out her dreams each year"?

Who used relative relationships, such as "Edgardo's dream salary of $100,000 would pay more than twice as much as Adeebah's ($45,000) each year" or "Adeebah's dream salary pays 45% of what Edgardo's would"?

Summarize:

So you could compare in different ways: by telling how far apart the amounts (the differences) are or by looking at the relative relationship between the two amounts using multiplication or division.

In this unit, comparing relative relationships will be an important mathematical idea.

Activity 2: Initial Assessment

Begin the *Initial Assessment* with a review of the questions. It is recommended that you display the problems on an overhead projector. In addition, distribute a copy of the *Initial Assessment* to every student. As a volunteer reads each question, ask the class:

How many of you think you can solve this problem?

How many of you think you cannot do it?

How many of you are not sure you can do it?

As the class reads and responds to each problem, keep the momentum going, moving onto the next problem after they record their judgments on their own copies. Tally responses on the *Class Tally* (*Appendices*, p. 150).

The *Initial Assessment* is a flexible tool. Depending on the class responses, you may decide to assign some or all of the problems. You might have the students work through the tasks, attempting the ones they feel most confident about first. Once they have completed as many as they can, discuss whether their first predictions about their ability to solve the problems changed. If so, why? Be sure to collect all of their work to analyze.

See the *Initial Assessment Checklist* for information on instructional decisions.

Activity 3: Mind Map

Start the Mind Map by writing "Rate, Ratio, Proportion" in the center circle. Ask:

What comes to mind when you think of those words?

Take a few ideas. Model clustering them, linking related ideas with lines. Then ask students to record their own ideas in the *Student Book*, p. 4. After a few minutes, ask students to compose one or two ideas and add them to the class Mind Map. Invite them to create new clusters as needed. Save the class Mind Map to compare with another the class will complete in the final session of the unit.

Summary Discussion

Orient students to their *Student Book* and review the *Unit Goals* in the *Student Book*, p. 5. Ask:

What do you understand the unit goals to be?

Can you think of anything else you need or want to know?

Direct students to add any new vocabulary words introduced today into the *Vocabulary* section of their *Student Book*, pp. 139–140, and write in the *Reflections* (*Student Book*, pp. 141–144). Also, after reviewing the *Unit Goals*, students might add goals of their own.

Looking Closely

Observe whether students are able to

Demonstrate ways to compare amounts, either additively (e.g., $300 is $225 more than $75) or multiplicatively (e.g., $300 is four times as much as $75, or $75 is 25% of $300)

Students may compare either way. Observe the ways they compare. If they only compare additively, encourage them to think of other comparisons, using fractions or percents or stating how many times larger one amount is than another. Proportions are grounded in relative reasoning, so you want to find out early whether students are already doing this. If many students have trouble using multiples, fractions, or decimals to compare amounts, consider returning to the *EMPower* units *Using Benchmarks: Fractions and Operations* and *Split It Up: More Fractions, Decimals, and Percents.*

Demonstrate ability to solve various proportional and nonproportional problems and describe solution strategies

One of the objectives of this opening lesson is to discover the strategies students already use when thinking proportionately. Rather than suggest they use a particular strategy, encourage them to think about how they usually try to solve similar situations on their own.

Pay attention to the types of strategies and tools students use. Notice whether they have created a diagram to figure out another equal ratio or whether they scaled up or down by using multiplication or division. Did students use unit rate for each problem? Or did they try to use the cross-product property when they were confronted with three of four numbers? Did students vary their strategy based on the type of number used or the specific situation?

Given a rate or ratio, can students project to or predict an equivalent ratio? Can students scale up or down by doubling or tripling to create a new set of ratios? Or do they add or subtract the same amount from each number in the ratio? Students who have difficulty using multiplicative reasoning will need more support in this unit.

Given three quantities, are students able to find a fourth in a proportional situation? Can students see the multiplicative relationship between the numbers in the first ratio? Can they see the relationship across one of the sets of numbers? This unit will give students opportunities to actually see the multiplicative relationships within and across sets of ratios.

How comfortable are students with the less friendly numbers and relationships?

Share experiences with rate, ratio, and proportion

The Mind Map will provide some insight into students' previous understanding of and attitudes toward ratio, rate, and proportion. Some students' understanding will be more developed than others; these students might refer to "cross product" or "factor." Other students might provide examples related to rate, such as "rate of speed," "tax rate," "mortgage rate," or "student:teacher ratio," but may be unable to think of ideas for the term "proportion."

Understand the scope of the unit

This assessment session ends with a presentation of the *Unit Goals* and asks students to state their own goals. Use their responses to inform your instructional goals for this unit.

WHAT TO LOOK FOR IN *OPENING THE UNIT*	WHO STANDS OUT? (LIST STUDENTS' INITIALS)			NOTES FOR NEXT STEPS
	STRONG	ADEQUATE	NEEDS WORK	
Concept Development • Accurately compares amounts with absolute differences • Accurately compares amounts with relative relationships				
Expressive Capacity • Articulates personal dreams and can predict the amount of income needed to fulfill those dreams • Offers ideas related to the terms "rate," "ratio," and "proportion" • Self-assesses strengths and weaknesses (*Initial Assessment*)				
Use of Tools • Uses diagrams, multiplication/division (scaling up or down), unit rates, cross product or other strategies to show that two ratios are equal • Uses effective strategies to solve problems				
Background Knowledge • Draws upon prior experience with ratio and proportion				

Rationale

This is an assessment session, an opportunity to find out how students reason about proportion. Because proportional reasoning is so commonly used in daily situations, students may already have a bank of strategies on which to rely. This assessment is designed to probe into students' thinking.

During the next weeks, students will develop the skills and conceptual understanding that enable them to solve problems in proportional, as well as nonproportional, situations. They will come to better understand how to use a variety of strategies and notations in numerous contexts to support their thinking, including using objects, diagrams, the property of equal fractions, tables, graphs, unit rates, and the cross-product property. As the unit progresses, they will be asked to deal with increasingly more complex numbers.

Although they may not be familiar with the terms "ratio" or "proportion," chances are they are familiar with the term "rate." This should be enough to get them to share their ideas and experiences related to the topic.

Math Background

Three mathematical ideas are touched upon in the opening activities:

1. Multiplicative relationships are a cornerstone of proportional thinking.
Activity 1 focuses on comparisons. Comparisons between any two quantities can be made with additive reasoning by expressing the absolute difference between the amounts (200 is 50 more than 150). Using multiplicative reasoning, comparisons can be made that focus on the relative relationship between the two quantities (200 is $1\frac{1}{3}$ as great as 150, or 150 is $\frac{3}{4}$ of 200). Although both are correct comparisons, the multiplicative or relative relationship is the cornerstone of proportional reasoning.

Research has shown that the numbers themselves and the nature of the relationship between the numbers affect the difficulty of the problem. For example, in Task 1 (*Activity 2*) shown here, 6 boxes are three times as many as 2 boxes, so a person might easily reason that the cost would be three times as much ($24). However, that $20 is $2\frac{1}{2}$ times as much as $8 might not be as easy to see.

2 boxes for $8

a. 6 boxes for ______

b. ______ boxes for $20

2. There are many strategies to use as one reasons about rates, ratio, and proportion.
Researchers who studied shoppers comparing prices have noticed that the relationship of the quantities often determines the strategy a person chooses. Unit pricing is not always used, nor is the school-taught process of cross-multiplication.

3. Some situations are proportional and others are not.
This important idea is probed in Task 4 (*Activity 2*), where the cost of the cardboard is not in the same relationship as the dimensions are to one another. Tripling the dimensions does not triple the area.

Context

The idea for *Activity 1* cropped up after reading about Juliet Schor's research on the income Americans feel they would need in order to fulfill their dreams. For insight (and lots of proportions), you might want to look at excerpts from the introduction in her first book on the subject, *The Overspent American: Why We Want What We Don't Need.*

Facilitation

If you display and read only one task at a time, you can ensure that students focus on just one situation at a time.

FACILITATING LESSON **1**

A Close Look at Supermarket Ads

Where do you see ratios used in supermarkets?

Synopsis

This lesson uses the familiar context of grocery ads to introduce the concept of **proportionality**. Students use pictures and numbers to reveal the multiplicative pattern that underlies equal ratios for grocery items sold by the pint, pound, item, etc.

1. The class identifies various ways ratios are expressed in supermarket ads.
2. Student pairs choose two ratios from a supermarket flyer, translate each into fraction form (*a*/*b*), and generate three equal ratios for each. They use diagrams or objects to show that the relationship *within* the ratios remains the same, even though the numbers change.
3. Students then look for numerical patterns *between* equal ratios and state mathematical rules for generating equal ratios.
4. Students create ads for bulk buying that have equivalent and nonequivalent relationships.
5. The class summarizes by restating the property of equal fractions.

Objectives

- Identify ratios in everyday consumer advertisements
- Determine equal ratios and demonstrate their equality with a diagram
- Develop a mathematical rule for establishing equal ratios
- Explain whether two ratios are equal by using a diagram or the mathematical rule (property of equal fractions)

Materials/Prep

- Game chips, pennies, or other objects to use as manipulatives
- Index cards or similar size scrap papers
- Local supermarket flyers (preferably brought in by students) or *Supermarket Flyers* (*Blackline Masters 1, 2,* and *3*), enough copies of each for every pair of students
- Scissors
- Tape

A Note on Vocabulary: Throughout the unit, introduce vocabulary by connecting it to students' everyday experiences. Students can develop a class vocabulary list of working definitions that they will refine as their understanding develops. The following are vocabulary words to highlight in *Lesson 1*:

A **ratio** is a way to compare two numbers or quantities. It can be written in fraction form (a/b), using the word "to" (a to b), or with a colon (a:b).

A **proportion** is a statement that two ratios are equal; for example, 1/3 = 2/6 is a proportion.

Per means "for each" or "for every."

Opening Discussion

Students begin by reflecting on their own personal experiences with how they make choices when they shop for groceries.

Start the discussion by saying:

Americans have so many choices when they go to the supermarket. What do you take into consideration when you buy an item?

On the board, record the students' criteria, for example:

- Favorite brand
- Quality
- What my kids will eat
- How much space I have for storage
- How much I am able to use before the food spoils
- Price
- Most for my money

Acknowledge the salience of the practical factors, but focus on the idea of getting the most for the money.

Ask:

What if you are very cost conscious and want to buy the cheapest product? When you are looking at an ad, what do you have to consider?

Two ideas should emerge: how much or many you get and how much it costs.

Distribute supermarket ads, either ones from local supermarkets or the *Supermarket Flyers* (*Blackline Masters 1, 2,* and *3*).

Ask:

What are the different ways an ad can tell how much you have to pay for a certain amount?

Be sure that examples similar to the following are mentioned:

- 2/$5
- 2 for $5
- $5 per bag
- 2 lb. for only $5.00
- 2 lb. $5.00

Tell students:

Sometimes the number or amount of the item purchased is understood as "one," as in the following cases:

- $5.00/pkg. = $5/1 package
- $5.00 ea. = $5/1 item
- $5.00 = $5.00/1 item

Summarize by making the following points:

These are different ways to express the relationship between two quantities. In mathematics, this relationship is called a "ratio."

Note that the fraction bar is often used as a signal to compare any two quantities or numbers. This meaning is different than "part out of a whole," which is the focus of the *EMPower* units *Using Benchmarks* and *Split It Up*. A fraction can be a signal for comparison of any two quantities, not necessarily a part:whole relationship, and this is the meaning to be studied in this unit.

Remind students to be attentive to labels. Write the following pairs of ratios on the board, and ask whether each pair means the same thing. Encourage students to draw diagrams to show their reasoning.

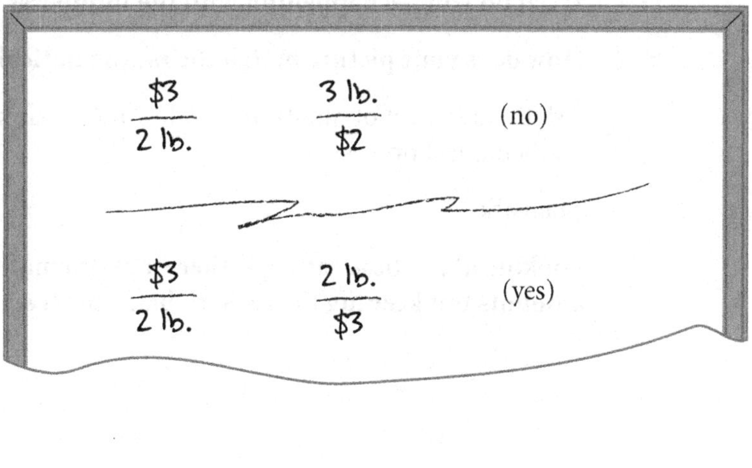

Activity 1: Different Numbers—Same Ratio

Refer students to *Different Numbers—Same Ratio* (*Student Book*, p. 8). Ask student pairs to choose any two ratios in the supermarket ads, cut them out, and tape them on pp. 8–9 of their *Student Book*. After writing the ratios they chose in fraction form, their task is to develop three other situations that have different numbers but keep the same ratios as the ones in their original ads. Encourage them to draw a picture that clearly shows that the deals are the same, even though the numbers change. When students are done, ask several volunteers to share the diagrams they made to show that the relationship between the quantities remains the same.

For example, someone might show that eight items for $12 is the same as two items for $3 by partitioning the 8:12 ratio into 2:3 groupings using a diagram such as the one below:

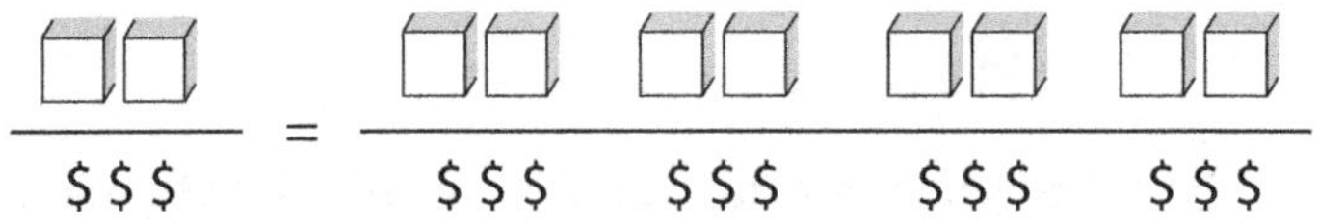

If students create a diagram such as the following one, push for meaning by asking where they see the ratio 2:3 in the diagram.

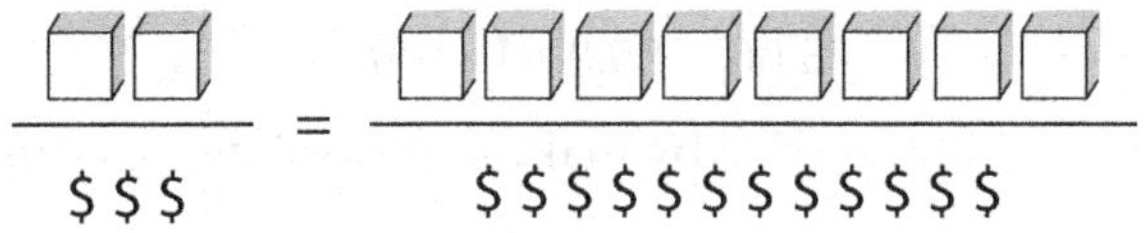

Activity 2: From Pictures to Mathematical Rules

Ask student pairs to examine one set of ratios from an ad from *Activity 1* to complete *From Pictures to Mathematical Rules* (*Student Book*, p. 10). After students have worked for a while, ask one pair to post their set of ratios, and discuss Problems 2–4.

What mathematical operation could you use to get from the first ratio to the new ratios?

What do you see happening with the numbers?

How does your picture match the mathematical pattern in the numbers?

Ask this same set of questions of a few more pairs, always probing for the mathematical process.

Then ask:

Looking at all these ratios, is there a mathematical rule for changing the amounts but keeping the ratio in balance? If so, what is it?

Some students may focus on the general growth:

- "You pay more to get more."

Some students may focus on additive patterns:

- "If you add another item, you add another amount of money."

Others may see multiplicative patterns:

- "If you double the amount, you double the cost."
- "Get 10 times more, pay 10 times as much."
- "By whatever number you multiply the amount, multiply the cost by the same number."

Ask students to write out their generalizations and then test them using the ratios from the ads they selected.

Heads Up!

Focus on the generalization: When you multiply or divide both amounts in a ratio by the same number, you will create another ratio that is equal to the first. In other words, the ratio relationship stays the same, even though the numbers change. Refer to this as "the property of equal fractions" or "the property of equal ratios."

Activity 3: Buying in Quantity

Refer students to *Buying in Quantity* (*Student Book*, p. 11). Students work in pairs to consider the pros and cons of buying in bulk. Once all pairs have had a chance to discuss this issue, have them share their ideas. Record their responses on the board.

Then close the discussion by posing questions such as the following:

What kinds of items can be bought in bulk? What items cannot be bought in bulk?

Is it always cheaper to buy in bulk? How do you know?

Now refer students to the next part of the activity. Explain that they are to select an ad from a flyer that they used in *Activity 1* and use it to create a set of ads for buying in bulk. Ask for a volunteer to read the directions:

- Select an ad from the flyer you used in *Activity 1*. Write out the original ad.
- Create a new ad that keeps the ratio the same so that the deal is exactly the same whether you buy in large amounts or not.
- Create a new ad in which it is cheaper to buy in bulk.
- Create a new ad in which it is more expensive to buy in bulk.

Distribute four index cards to each pair. They are to write each of the four ads they created on an index card. Tell them not to mark the cards with the identifying letters ("A," "B," "C," and "D") because they will share their completed cards with another pair of students.

Once each student pair has created their own set of four ratios, have them exchange their cards with another pair of students who then have to determine which ratios represent a better deal, the same deal, and a worse deal when buying in bulk. Encourage students to use visual representations or a mathematical rule to explain why they know which ratios are equal to the original and which are not.

Do pairs agree and can they explain their reasoning using pictures or a mathematical rule?

Summary Discussion

Review today's lesson by having students consider how they talked about ratios. Ask:

What terms did you use to talk about ratios?

Ask students to share their definition of the term "ratio." Add these terms and definitions to the class vocabulary list, and ask students to copy them into the *Vocabulary* section (*Student Book*, pp. 139–140).

Then ask:

If you double both the cost and the amount, will the ratio always be the same?

If you add the same number to both the cost and the amount, will the ratio always be the same?

Ask students for examples to support their reasoning.

Then have a volunteer summarize the property of equal fractions.

Direct students to *Reflections* (*Student Book*, pp. 141–144), where they write their own explanations for keeping ratios the same. Suggest that they also explain some situations when ratios do not stay the same .

Heads Up!

There is a lot of prep needed for the next lesson. Take a look at the list of materials for *Lesson 2* now so you have time to gather all the supplies.

Practice

Balancing Ratios, p. 13
Asks students to create new, but equal, ratios.

Using the Property of Equal Fractions, p. 15
Provides practice seeing the multiplicative relationship between two ratios.

Is Bigger Cheaper? p. 17
Gives students opportunities to reason about ratios without using numbers.

Working with Larger Numbers, p. 18
Offers work on larger whole numbers (multiples of 10).

Mental Math, p. 19
Provides practice for students to sharpen their mental math strategies.

Extension

Extension: Ratios in My Life, p. 21
Students broaden their focus of ratios to include situations other than shopping.

Test Practice

Test Practice, p. 22

Looking Closely

Observe whether students are able to

Identify ratios in everyday consumer advertisements

Do students pay attention to labels? Are they alert to the difference between two cans for $3 and $2 for three cans? Ads indicate the price per amount in different ways. Are students fluent with the various notations? Do they use the term "for every" interchangeably with "per"? What words do they use to interpret the fraction bar? Do they see the implied "1" in an advertisement such as "$4 ea."? It is important that students not only know how to recognize ratios, but that they also be aware of what two amounts are being compared in the ratio. If students struggle, encourage them to label their ratios so that what is being compared is clear.

Determine equal ratios and demonstrate their equality with a diagram

Are students able to draw a diagram to illustrate a ratio? Even if students know how to use cross products, they should be encouraged to draw the pattern so that they can actually visualize the relationship between the two amounts. Some students may have only memorized the rule for cross products but have not had the opportunity to explore the concept of ratio and proportion more deeply.

Seeing that there are 3 sets of 2 in 6 and 3 sets of 3 in 9 brings meaning to the proportion 2/3 = 6/9, much more so than cross-multiplying 2 × 9 and 3 × 6. The fact that you can change the numbers in the ad, yet maintain the balance between the cost and the number of items, is a big idea. Visualizing how the balance remains the same establishes a solid foundation for proportional reasoning.

When you encourage students to draw a picture to illustrate the set of ratios, it is important that they take this task seriously. It is more than just drawing the total number at the top and the total number on the bottom. They should be able to organize the totals into sets such as the following:

$\frac{2}{\$5} = \frac{8}{\$20} =$

$ $ $ $ $ $ $ $ $ $ $ $ $ $ $ $ $ $ $ $

In doing so, students not only see the pattern of 2 for every 5 within 8/20, but they also begin to notice that there are 4 groups of 2 for every 4 groups of 5, which grounds the mathematical rule.

Develop a mathematical rule for establishing equal ratios

Are students able to develop a mathematical pattern? Research has shown (and our experience in field tests supports this idea) that most people reason proportionally by using a combination of strategies, including doubling and halving. These strategies are most commonly used in real-life situations, especially when the numbers are friendly. Many adult students also use these strategies, and it is powerful for them to learn that it is acceptable to use them. People often scale up or down by holding the relationship constant and "building up" or "building down." For example, when asked to find how many items are being sold for $100, someone might build up from knowing "4 for $6," to 8 for $12, to 16 for $24, to 32 for $48, to 64 for $96 (almost there!). Or he or she may reason, "Four for $6 means 20 for $30, 60 for $90, and another $10 would buy about six more items for a total of 66 items."

If students have difficulty with creating a general rule, you will need to determine whether the problem is that they do not have a firm grasp of their multiplication tables (in which case you should ensure that they have a multiplication table chart and/or calculator handy) or that they cannot see the relationship between the sets of numbers. If the latter is the case, encourage students to use two colors of manipulatives or two different kinds of objects and separate them into sets. Once they are comfortable with the objects, encourage students to continue drawing the sets so that they can actually see the number of sets. This reinforces the concrete understanding of equal ratios that students need. They can then count the number of sets in the expanded ratio. Students who are struggling with articulating a rule should continue to work with small, friendly numbers until they are comfortable with the rule.

Students may connect the rule for arriving at equal ratios by multiplying or dividing both amounts with previous learning about renaming fractions. Whether one uses larger or smaller fractions when renaming the fraction, mathematically the action is the same:

$$\frac{a}{b} = \frac{a}{b} \text{ x } \frac{c}{c}$$

Explain whether two ratios are equal by using a diagram or the mathematical rule (property of equal fractions)

How do students determine whether two ratios are equal? It is important to direct them to look carefully to determine whether sets of ratios are equal. Students should be able to use a visual pattern to show equality or inequality of ratios. Allowing students to explore and develop their own strategies for reasoning about ratios helps them really "see" the ratios. This is one student's method to determine whether a set of ratios was eqivalent to 2/3.

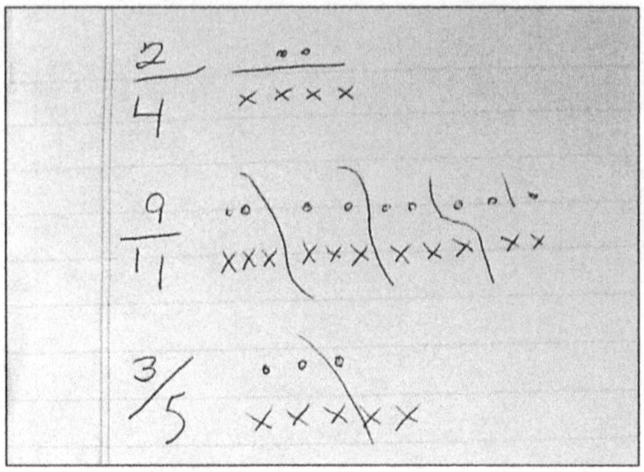

WHAT TO LOOK FOR IN *LESSON 1*	WHO STANDS OUT? (LIST STUDENTS' INITIALS)			NOTES FOR NEXT STEPS
	STRONG	ADEQUATE	NEEDS WORK	
Concept Development • Identifies ratios in supermarket ads • Understands when two ratios are equal or not equal • Compares supermarket ratios to distinguish a good deal from a bad deal				
Expressive Capacity • Translates a supermarket ad into a ratio in fractional form • Writes a ratio in several forms (*a*/*b*, *a*:*b*, *a* per *b*) • Can explain clearly whether two ratios are equal				
Use of Tools • Uses pictures or objects to show that two ratios are or are not equal • Uses a mathematical rule (such as "multiply both amounts by the same number") to create a new ratio equal to a given ratio				
Background Knowledge • Draws upon shopping savvy				

Rationale

Students who are able to visualize equal ratios have an advantage over those who are only able to use cross product or unit rate as strategies for working with proportional problems. "Seeing" the relationships brings meaning to what students may construe as decontextualized number-crunching. Pictures and diagrams help trace how ratios stay in balance even though one of the amounts may shrink or grow.

By the end of this unit, students should be comfortable using more efficient strategies than drawing patterns, but the time spent up front in the first few lessons helps students build a more robust understanding of the concept of equal ratios.

Math Background

In this lesson, important mathematical concepts about proportion are introduced.

Multiplicative relationships abound in proportional situations

Examining the relationship between numbers—comparing numerical values—is a central activity in mathematics. There are two ways to compare numbers: additively and multiplicatively. Take, for example, the numbers 1,000 and 250. One could say 1,000 is 750 more than 250. This is an additive comparison, where the operation of addition or subtraction applies. On the other hand, a multiplicative comparison leads to the conclusion that 1,000 is 4 times as great as 250, 250 is ¼ of 1,000, or 250 is 25% of 1,000. The operations of multiplication and division support this comparison. Both comparisons—additive and multiplicative—are mathematically correct, but proportional reasoning relies on multiplicative comparison.

Rates and ratios are communicated in a variety of ways

It is essential for people to be fluent with the written language of proportion. In everyday situations, these relationships come at us in many forms: with a slanted fraction bar (10/$3); the word "for" (10 for $3); or with no signal other than the amount and price (10 lb. $3). "Each" and "ea." imply "for one" ($3 ea.). Less common in everyday situations, but frequently used in math books, is the colon (3:1).

In this lesson, students are encouraged to use the fraction bar as a signal for comparing one amount to another. This is a more general use of the fraction bar than a signal for the part/whole relationship developed in *Using Benchmarks* and *Split It Up*. The fraction bar in *Keeping Things in Proportion* is used for a comparison of any numbers or amounts: part/whole, part/part, or like units to like units.

Fractions and ratios share significant mathematical properties when considering equality

Although the meaning of a fraction is broadened, ratios "act" like fractions in some ways:

Part/whole fractions and ratios share the property that when you multiply or divide both parts, equality is preserved. As one student said, "Double one, then double the other."

$$\frac{6}{8} = \frac{6}{8} \times \frac{2}{2} = \frac{12}{16}$$

In this lesson, students are asked repeatedly to show with pictures how the relationship is preserved, even though the numbers change. Buying 6 cans of soup for $8 is essentially the same deal as buying 12 cans for $16. Of course, the actual amounts are not the same, but the multiplicative relationship between them is. They connect their pictures with the mathematical rule, which in more formal mathematics is the multiplicative property of one, that is, $a/b = a/b \times 1 = a/b \times c/c$.

In some cases, ratios do not "act" like fractions. For example, ¼ of a dollar + ½ of a dollar = ¾ of a dollar has meaning when adding part/whole fractions, i.e., each fraction is part of a whole dollar. On the other hand, it does not make sense to add the ratios ¼ lb. of ground beef and ½ lb. of ground turkey for a total of 2⁄6 lb. of ground beef turkey.

Context

Researchers (of children's mathematical thinking) have found that familiarity of context contributes to the relative ease with which people are able to think proportionally (Lamon, 1993). This finding guided the choice of grocery shopping as the context with which to launch the unit. It is hard to think of a context more familiar to adults than shopping. Adults can draw on experiences in which they used proportional reasoning, such as comparing prices. They frequently need to juggle the relationship of two pieces of information: the given price and the quantity for that price. Savvy shoppers—those who attend to prices—have a keen sense of when they are getting the most for their money.

Observations of adult grocery shoppers in actual shopping situations indicate that shoppers use many strategies to reason well about which of two deals is the better buy, and those strategies depend on the nature of the numbers and the relationship of those numbers to one another (Lave, 1989). Although many mathematics textbooks encourage unit pricing as the correct way to approach each of these situations (in the example below, finding the price per ounce), that is not the way real shoppers normally do their math in the supermarket.

People tend to reason differently when comparing, for example,

> 12 oz. for $0.89 or 12 oz. for $1.09;
>
> 12 oz. for $0.89 or 24 oz. for $1.50; and
>
> 12 oz. for $0.89 or 20 oz. for $1.50.

In the first situation, the quantity of the item is the same, so one merely needs to consider the money amount ($0.89 is less than $1.09). In the second instance, noticing that the second quantity is double the first would confirm that one must double the cost of the first, smaller amount to make the comparison. In the third situation, finding the unit price might be the most efficient strategy, but a person with good number sense might take another route, reasoning 4 oz. of both would cost $0.30.

This unit slowly builds from comfortable numbers to larger, more challenging numbers. It also includes not only whole number relationships, such as doubling and tripling, but also benchmark fractions, such as 1½ times an amount.

Facilitation

Students will only be as comfortable creating visual representations or using manipulatives as you are, so it is important that you also become comfortable working with visuals. Students who feel drawing is childish may want to do simple illustrations using "X's" and "O's."

Students who are not comfortable using calculators may need a quick review on how to determine the rule for establishing equal ratios.

Making the Lesson Easier

Spend as much time as needed on visual representations and graphic models, if students are struggling. Work with friendly whole numbers only so that students can draw the different equal ratios. Representing the situation using chips, coins, square-inch tiles, or other manipulatives is helpful.

Making the Lesson Harder

If students are comfortable with the cross-product strategy and think drawing patterns is a waste of time, challenge them to consider how to represent graphically such proportions as 0.5 lb./$0.75 = 6 lb./$9.00.

When she asked students to begin to generalize about equal ratios, one teacher elicited these responses:

Bob told me, "Ratios getting larger or smaller are just patterns. If one grows, they both do. If one shrinks, they both do."

Ellen agreed. She said, "If you can divide the two, you know you can figure out the others. So it's just raising or reducing fractions."

Terri hates fractions. She said she knew what a ratio was, but it seemed hard without a picture.

Starting with a supermarket ad, this student recorded his thinking about equivalent ratios and showed good use of the fraction bar:

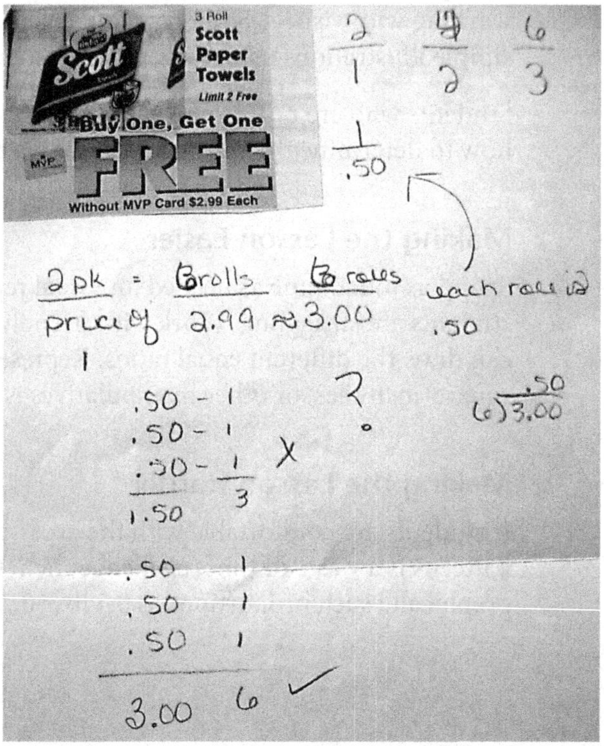

Lisa Mullins
Hawkins County Adult Education Program, Rogersville, TN

Students often connect and verify previous learning during class discussions.

"Does order matter?" asked one student, as she set up her ratios.

Another student responded, "I'm not sure of the order. It seems that somewhere, far back in my head, I was told that you had to write the ratio in a particular order. So, do I write '2 for $5' as 2/5 or 5/2?"

I explained that either way worked as long as we knew what the "2" and the "5" represented. Then we proceeded to use chips to prove the point. This worked well because another student was sitting opposite her. When she put two yellow chips over five white chips, she saw 2/5 although he saw 5/2. She could see that her 4/10 would be the same as 10/4 for him. In both cases, the pattern was consistent.

Donna Curry
Gardiner Adult Education Program, Gardiner, ME

FACILITATING LESSON **2**

It's a Lot of Work!

How does taking a sample help in predicting?

Synopsis

This lesson extends the notion of rate, drawing attention to the capacity of ratios to help extrapolate from smaller amounts to larger ones. Students use ratios to predict how long it will take to perform a larger task based on the time it took to conduct a small portion of that task. They explore how the underlying ratio between the two quantities in the small sample guides estimation for larger quantities.

1. The class discusses ways to predict how long it will take to complete tasks at work or home.
2. In teams, students estimate how long it will take to accomplish three tasks based on the time it takes them to do a portion of each task.
3. After all estimates are displayed and methods discussed, the class agrees on a final estimated time it would take to complete all three tasks of the assignment.
4. The class lists advantages for performing the sample task more than once.

Objectives

- Conduct and describe a sample of work over a period of time
- Use the sample to make a prediction for a larger amount by reasoning with equal ratios

Photo by RODNAE Productions from Pexels

Materials/Prep

You will need to set up three kinds of stations before the class begins. If the class is large, set up multiples of each station.

Station 1

Stuffing Envelopes: a pile of 8½" x 11" paper (about 10 sheets per group) and about 10 envelopes. Fold one in a trifold and stuff it in the envelope as an example.

Station 2

Potato Salad: one medium size potato (weighing about ½ lb.) for each team that will use this station and a potato peeler.

Station 3

Rolling Pennies: a jar of pennies—no more than 200—and a few empty penny rolls.

Prepare the following chart to display during class, fill in, and save for *Lesson 7*:

Task	Time Estimates		
	Team 1	Team 2	Team 3
1. Folding and stuffing 1,000 envelopes			
2. Peeling 50 lbs. of potatoes			
3. Rolling 10,000 pennies			
Total Time			

Opening Discussion

Open the class by asking students to share their methods for estimating how much time it takes them to complete tasks at work or home. They might use experience as their guide and say something such as,

"I always know it's going to take me half an hour to answer my e-mail."

"Twenty minutes to make my children's lunches everyday."

Or they might say that it depends on what the work load is that day. For example, if there are 10 people for dinner, it takes longer to wash the dishes than if only four people have had dinner. If students bring up this issue, ask:

How can you figure out how much longer the task would take?

Ask if anyone has had a job where they had to make predictions about how long it would take to complete a large amount of work. Listen for sampling ("In housekeeping, I know it takes me about 45 minutes to clean a room, so I should be able to do 10 rooms a day.").

Explain that today they are going to take sample rates of work in order to predict how long bigger tasks would take.

Activity: How Long Will It Take?

Have students read silently or aloud *Activity: How Long Will It Take? (Student Book*, p. 24).

Teams Travel to the Stations

Break the class up into teams of two to four. Each group needs to start at a different station, but all groups travel to each of the three stations before they are done.

Remind teams that they have 40 minutes to come up with a decision on how long the volunteer job would probably take, and they need to have at least two ways to verify their estimates.

They should complete pp. 25–28 (*Student Book*).

As you circulate, notice strategies. The numbers may get unwieldy, so encourage students to make the numbers friendlier by rounding or changing time units.

Ask teams to post their estimates for each of the jobs on a table you have prepared.

Pulling the Class Together

Pull the class together to comment on the strategies used to arrive at their answers and the range of answers that were determined.

Ask for a few sharings, such as these:

How did Team 2 arrive at the time it would take to fold the flyers and stuff the envelopes?

Could you show that another way with a picture or equal fractions?

Now ask for people to comment on the range of answers:

Where are we most in agreement?

What sample rates did teams use?

What can you say about those rates?

Where is the biggest discrepancy?

Next ask people to consider:

If we needed to agree on one estimate for each job, what do you think that would be?

This could force the issue of **averaging**, or students might use other strategies. For example, the class might pick the most common estimate (mode) or the center estimate (median). They may need to think about what to do if there is one estimate that is very different from the rest of the class's estimates.

Summary Discussion

Ask:

How did ratios help you predict the time to complete a large task?

How would you predict the time needed to stuff 100,000 envelopes? To peel 250 pounds of potatoes?

What are other situations where sampling and figuring out ratios for a smaller task or a shorter time period might help?

Some people say that it is always good to sample more than once. What do you think the advantages and disadvantages are of doing so?

Then suggest that students think about what they have learned today by responding in the *Reflections* section (*Student Book*, pp. 141–144).

Heads Up!

Save the class chart of predictions. You will refer to it again in *Lesson 7,* when students do yet another type of sampling.

Heads Up!

There is a lot of prep needed for the next lesson. Take a look at the list of materials for *Lesson 3* now so you have time to gather all the supplies.

Practice

Time to Be Proportional, p. 29
Provides practice using mental math for converting between units of time.

Wasting Water, p. 30
Gives students further practice taking samples, then predicting for larger amounts.

Typing Tests, p. 31
Offers another opportunity to predict from a sample and to compare rates.

Who Is the Fastest? p. 32
Offers practice comparing rates of speed.

Test Practice

Test Practice, p. 33

Looking Closely

Observe whether students are able to

Conduct and describe a sample of work over a period of time

Do students know how to take a sample? Do they consider what sample size might be reasonable? Allow students to determine for themselves how to take a sample. They should be able to figure out their sample size or the length of time they want to sample. Provide as little direction as possible, but offer guiding questions as needed.

Once students take a sample, listen actively as they figure out how to articulate their rate. For example, do they try to find how many potatoes are peeled in one minute? Or do they time how long it takes to peel one potato? Or do they settle on a rate with multiple minutes or potatoes? See *Lesson 2 in Action,* p. 35, for one way students struggled with setting a sample rate.

Use the sample to make a prediction for a larger amount by reasoning with equal ratios

Are students able to see how ratios can be used to predict for larger samples? Students should be encouraged to use pictures or to look for simple patterns from which to build to larger amounts. If the numbers are too unwieldy, encourage students to round to the nearest whole number or to work with seconds instead of parts of minutes.

What tools do students use to go from their sample rate to the larger numbers? In *Lesson 1*, students used pictures and/or a mathematical property of equal ratios—multiplying both amounts by the same number—to keep the rates or ratios equal. Another tool you or students might introduce is a table. In *Lesson 2 in Action*, p. 36, a table seemed to illuminate the relationship between seconds and minutes.

WHAT TO LOOK FOR IN *LESSON 2*	WHO STANDS OUT? (LIST STUDENTS' INITIALS)			NOTES FOR NEXT STEPS
	STRONG	ADEQUATE	NEEDS WORK	
Concept Development • Chooses a work sample and uses it to predict the amount of time for a larger amount of work • Maintains the relationship between the amount of work and the time taken				
Expressive Capacity • Translates a work sample into a ratio in fractional form • Writes a ratio in several forms (*a*/*b*, *a*:*b*, *a* per *b*) • Can explain clearly whether two ratios are equal				
Use of Tools • Uses pictures or objects to show that two ratios are or are not equal • Uses the property of equal fractions (multiplies both amounts by the same number) to create a ratio equal to a given ratio • Uses a table to keep track of the two amounts as they increase				
Background Knowledge • Draws upon knowledge of piece work or sampling in workplace situations				

Rationale

Estimating and making predictions from small tasks are integral to planning and accomplishing large tasks. Businesses use sampling techniques to predict the quality of their products and services, as well as to estimate the quantities needed.

Math Background

The ratio of the quantity of work done to the amount of time required is assumed to be constant in the situations at the stations. In this second lesson, students keep the rate constant. Their pictures, tables, or the property of equal fractions should reflect this. As in *Lesson 1*, the central mathematical ideas of *Lesson 2* are as follows:

1. The multiplicative relationship between the amount of work done and the amount of time is always the same. In other words, the relationship *within* the ratio remains constant.

   For example, when

   $$\frac{\text{1 potato}}{\text{40 sec.}} = \frac{\text{2 potatoes}}{\text{80 sec.}} = \frac{\text{3 potatoes}}{\text{120 sec.}} = \frac{\text{90 potatoes}}{\text{3,600 sec.}}$$

   the number of seconds is always forty times the number of potatoes.

2. An equivalent rate or ratio occurs when both amounts of the ratio are multiplied by the same number. Thus the relationship *between* any two equal ratios can be seen as a multiplication of both amounts by the same number to get from one to the other.

Rates often have to do with situations where the time varies with another variable, but all rates are not constant. When a person gets more experienced at a given task, his or her rate might speed up; or if the person is tired at the end of the day, his or her rate might decrease. Mathematically, situations where the rate changes do not lend themselves to proportional reasoning.

It is important to know that the term "sample" is not used in the statistical sense in this lesson, but in the everyday sense of sample—a portion, part, or segment.

Context

Just as in the first lesson, a familiar context is at the core of this activity. Time is money, especially for piece work. Although the tasks in this lesson are set in a volunteer context, students may have had jobs where they were paid by the unit of work completed or where they have been expected to complete a given number of tasks in a certain period of time.

Facilitation

This activity should be as student-focused as possible. Listen and allow teams of students to struggle among themselves to figure out how to address the challenges before them. Be sure to have calculators and manipulatives handy, but allow students to figure out for themselves the strategies they are going to use. Encourage mental math.

Do not push for students to establish unit rate, especially if the numbers are unwieldy. As noted in *Lesson 2 in Action*, p. 35, students tried to come up with a unit rate (1 potato = 40 seconds or 0.667 minutes). This was uncomfortable for them, so they established a more comfortable ratio (3 potatoes = 2 minutes). Consider whether the unit rate is actually going to clarify the situation for students before suggesting that they use it.

Making the Lesson Easier

If students find that they have chosen unwieldy numbers in their sample, you may want to stop them before they get too far with building on their ratio. Suggest that they round their ratio so that the numbers are friendlier. For example, in one class, the students predicted that it would take 65 seconds to stuff 6 envelopes. Even though many teachers would suggest rounding 65 to 70, it made more sense for students to round to 60 (because 60 is 10 times 6). This rounding helps students deal more readily with the process of building to larger ratios, yet doesn't impact the integrity of the sample.

Making the Lesson Harder

Ask students to make predictions for a given amount of time (as opposed to a given number of envelopes or potatoes peeled): "So, if someone did this for two days, taking into account eight hours of time set aside to sleep, how many potatoes could he or she peel?"

Ask students to revisit their initial sampling rates to see how changing the time unit would affect the outcome. For example, if a group begins with peeling one potato in 40 seconds, what would that time be in minutes? (2/3 min.)

$$\frac{\text{1 potato}}{\text{2/3 min.}} = \frac{\text{2 potatoes}}{\text{1 1/3 min.}}$$

Students take various routes as they negotiate through the stations. The size of the sample and the way they build up from that sample to larger amounts is their choice, as is their selection of a second strategy. There is a lot of room to "play with the numbers," and if they get muddled in the numbers, it is important to seek other strategies, such as creating a table. Here is one teacher's description of how two women, Carla and Maya, worked together and of the teacher's interventions.

According to their timed observation, it took 40 seconds to peel one potato. Maya grabbed a calculator to determine how many seconds it would take to peel 100, while Carla began her calculations using pencil and paper. They agreed that it would take 4,000 seconds.

I listened as they both tried different types of calculations. They divided 40 into 60 and got 1.5. They then tried [dividing] 60 into 40 and got 0.667. They didn't like either number, but they thought 0.667 was particularly useless. They divided 100 into 20 and got 0.2, which they thought might give them the number of minutes left "to do the rest of the chores [stations]." They didn't like 0.2; they felt it wasn't the right answer.

After the students had made many false starts, I finally decided that we might try to make a chart up on the board. Because Carla had been developing a vertical chart, we continued with her format. We started building up seconds and potatoes, actually adding on 40 seconds for every new potato.

Potatoes	Seconds
1	40
2	80
3	120

When we got to 3 potatoes and 120 seconds (40 × 3), Carla realized that there were 2 minutes in 120 seconds. This was a nice breakthrough for the pair.

continued on next page

continued from previous page

Because I wanted them to see this in fraction form as well, I suggested setting up the ratio (fraction form) of 3 potatoes to 2 minutes, and we worked together to get to 12 potatoes in 8 minutes. Maya could readily see the pattern and started doubling each amount in her head. She suggested that the next ratio would be 24 to 16. Maya and Carla worked together, mostly doubling, but also combining.

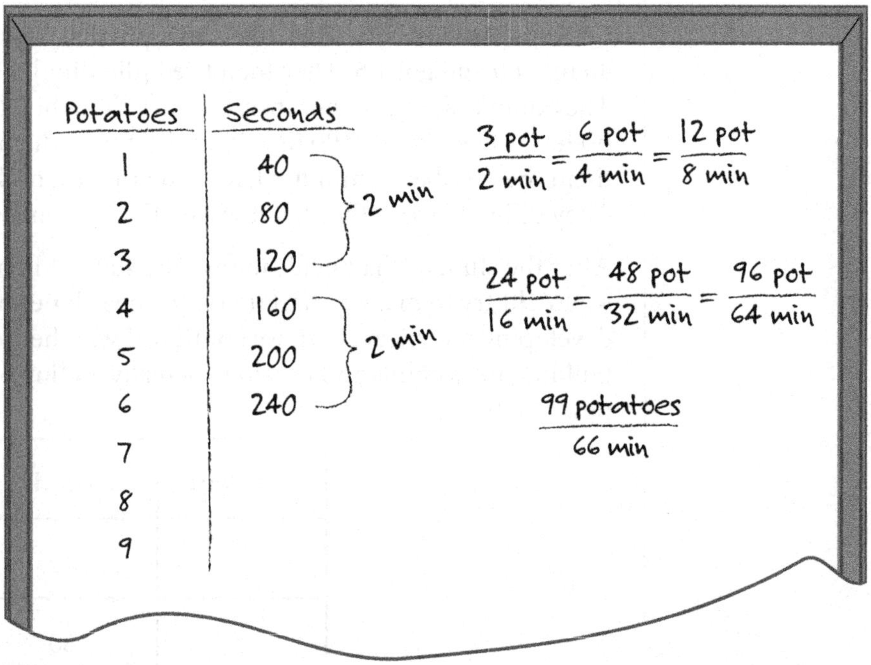

Building this way, they were able to get to 99 potatoes in 66 minutes. They knew this was close to 100 potatoes, so I asked what would happen to the time if they added one more potato. They agreed that it would add another 40 seconds on to the 66 minutes, for a total of almost 67 minutes. I then asked how many hours 66 or 67 minutes would represent. Carla estimated that it would take "1 hour and 6 minutes."

I would say that my role as a teacher was to encourage the two women to stay as close to the real situation as they could, visualizing the balance between potatoes and minutes. Their scramble for numbers on the calculator seemed to disconnect them from the meaning.

Donna Curry
Bailey Evening School, Bath, Maine

FACILITATING LESSON 3

Tasty Ratios

What do ratios have to do with taste?

Synopsis

This lesson gives students an opportunity to play with ratios using their sense of taste. The focus for students is twofold: to maintain the ratio of the ingredients when making a new recipe and to adjust ingredients in order to fix a failed recipe.

1. The class discusses the need for using correct proportions when cooking.
2. In small groups, students taste three orange juice mixtures and speculate about the ratio of water to concentrate. They compare their conclusions before checking them against the recipes actually used.
3. Students consider how to fix the failed mixtures so the ingredients are in the right proportion to one another.
4. Students summarize ways to ensure that the ratio of two ingredients is maintained.

Objectives

- Use ratios to describe taste and visual comparisons
- Determine how to adjust ingredients so that the proportions are correct

Photo by Kampus Production from Pexels

Materials/Prep

- Markers
- Easel paper
- Sticky notes
- Small paper cups (three per student) for sampling orange juice
- Three spoons for stirring mixtures

Prepare student copies of *Orange Juice Recipe* (*Blackline Master 4*) or *Alternate Orange Juice Recipe* (*Blackline Master 5*), depending on the number of students in your class.

Prepare to display, have students add and comment on, and save three tables like the one below—one table for each orange juice mixture (A, B, and C) in *Activity 1*:

Mixture A
Describe the taste:
Describe the look:
Guess the ratio of water to frozen concentrate:

Orange Juice Recipe for a class of more than five students

- 1 gallon of water
- 2 (12 fl. oz.) cans of frozen orange juice concentrate that calls for three cans of water to one can of concentrate
- 3 empty containers (at least one-half gallon capacity each) to hold the three orange juice mixtures

Prepare Mixture A, where the proportions are correct (3 parts water:1 part concentrate). Add three cans of water to one can of orange juice concentrate and stir. This will yield a normal mixture.

Step 1

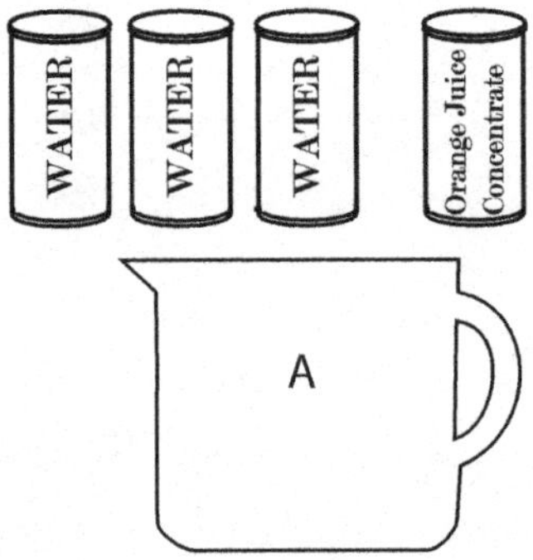

3:1

Prepare Mixture B with one-third the recommended amount of water (1 part water:1 part concentrate). Add one can of water to one can of orange juice concentrate and stir. This will yield a thick mixture.

Step 2

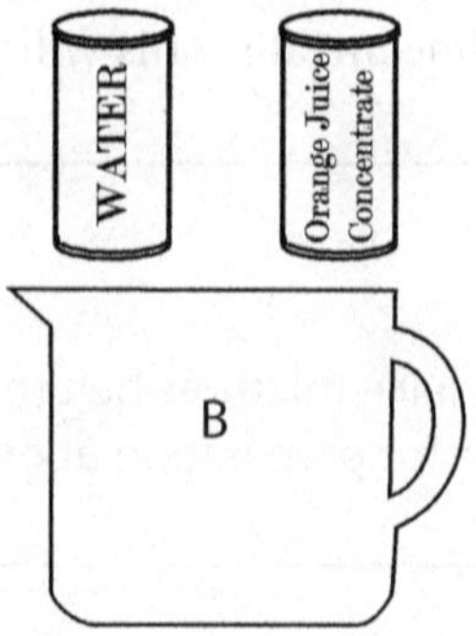

1:1

Prepare Mixture C, in which there is much more water than there should be (7 parts water:1 part concentrate). Pour 16 fl. oz. (2 cups—*not* cans) of Mixture A into a container and add 16 fl. oz. of water. This will yield a very weak mixture.

Step 3

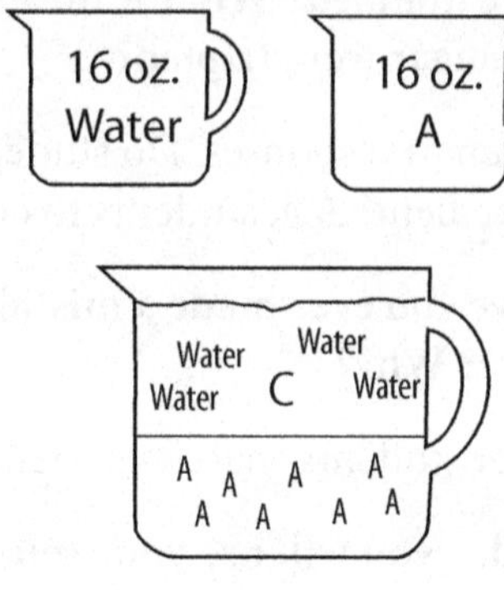

7:1

Alternate Recipe (for a class of five or fewer students)

- 1 (12 fl. oz.) can of frozen orange juice concentrate that calls for three cans of water to one can of concentrate
- ½ gallon of water
- 3 empty containers (at least one-half gallon capacity each) to hold the three orange juice mixtures

Divide the concentrate into thirds.

Prepare Mixture A, where the proportions are correct (3 parts water:1 part concentrate). Add one can of water to one-third of a can of orange juice concentrate and stir. This will yield a normal mixture.

Prepare Mixture B with one-third the recommended amount of water (1 part water:1 part concentrate). Add one-third of a can of water to one-third of a can of orange juice concentrate and stir. This will yield a thick mixture.

Prepare Mixture C, in which there is much more water than there should be (7 parts water:1 part concentrate). Add 2⅓ cans of water to ⅓ of a can of orange juice concentrate. This will yield a very weak mixture.

Heads Up!

Prepare orange juice mixtures before the class begins. Be sure to use the ratios described in the directions for preparation above.

Opening Discussion

Begin by asking students to consider the role of proportion in cooking. Refer students to the recipe for pancakes on p. 35 of the *Student Book*. Once they have examined the recipe, ask:

If only one of the ingredient amounts in this pancake recipe were changed, what would the pancakes taste like? For example, what if the amount of flour were doubled? What if the amount of baking powder were cut in half? What if the sugar were tripled?

Listen to responses. Do students have a sense of the effect of changing one ingredient? Ask students to consider their own personal experiences with recipes.

Have you ever made a mistake following a recipe? What happened? How did it taste? Why?

After students share experiences, say:

Today you will use your senses of taste and sight to make judgments about whether I used the correct proportions when making mixtures of orange juice.

Later you will figure out how to fix any mistakes I might have made.

Activity 1: Orange Juice Taste Test

For this activity, students will examine and taste a little of each of the three containers of orange juice that you prepared. Each container will be a different concentration, stirred and well labeled (A, B, or C).

Set up a station as shown in the picture (or pass around cups labeled A, B, and C). Also have either the empty can or the label available for students to examine.

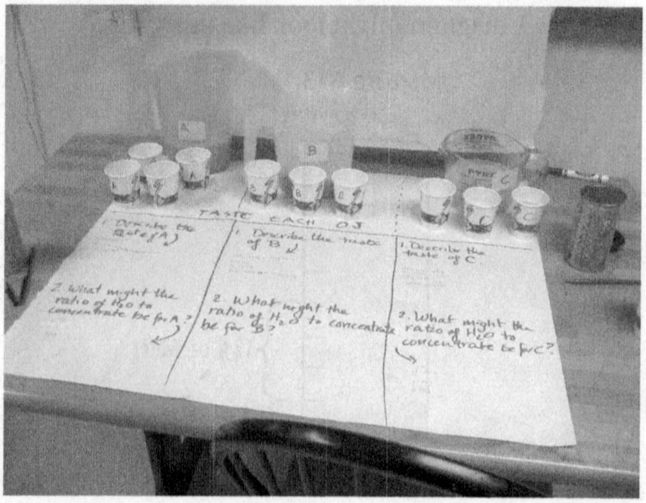

Ask everyone to pour a little of each mixture into three cups, labeling their cups "A," "B," and "C." Before taste-testing each mixture, ask students to look at the mixtures and predict what they think the ratio of water to concentrate is for each. Then ask students to taste a little of each to adjust their predicted ratio and then to write their predictions down on p. 36 in the *Student Book*.

Also ask students to share their predictions on a sticky note on the each of the displays "Mixture A," "Mixture B," and "Mixture C."

After everyone has written their responses, hand out your actual recipe (either *Blackline Master 4* or 5, depending on which recipe you used). Ask:

What do you think of the guesses? Which guess was the closest in each case?

Give students plenty of time to figure this out alone and in pairs, and then ask for responses. Figuring out Mixture C is a real challenge and should elicit discussion and a lot of drawing.

There are many ways to think about this. Some people might think about Mixture C this way:

> 16 fl. oz. of Mixture A has 3 parts water and 1 part concentrate, but Mixture C has 4 more parts of water. That makes 7 parts water and 1 part concentrate (14 fl. oz. of water to 2 fl. oz. of concentrate). So Mixture C is a 7:1 ratio.

Other students may interpret the Mixture C recipe as follows:

> 16 fl. oz. of Mixture A has 12 fl. oz. of water and 4 fl. oz. of concentrate (3 to 1). If you add 16 more fl. oz. of water, you get a mixture with 28 fl. oz. of water and 4 fl. oz. of concentrate. This is a 7 to 1 ratio of water to concentrate.

A diagram might look like this:

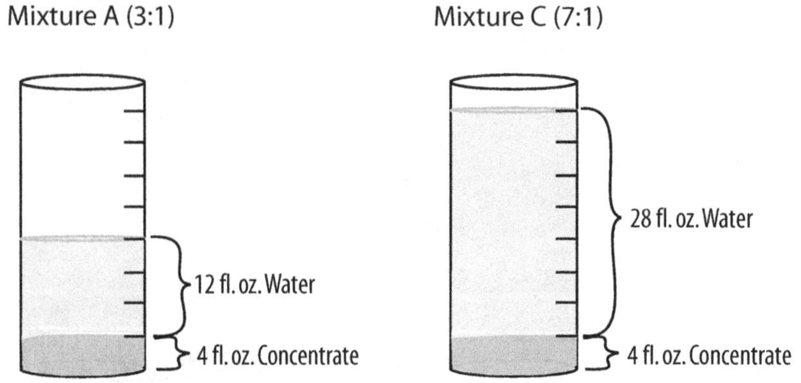

Activity 2: "Doctoring" the Orange Juice

Encourage students to work in pairs or triads to figure out how to create mixtures that are in balance from the failed Mixtures B and C.

Ask:

How would you fix Mixture B without throwing any away?

How would you fix Mixture C without throwing any away?

Encourage picture drawing.

For Mixture B, the "fix" is fairly easy if you add 2 more cans of water.

For Mixture C, you would have to add some more concentrate and possibly a little water. But how much? Some people might think of it this way:

If you have 28 fl. oz. of water and 4 fl. oz. of concentrate, and you want to bring the strength up to a 3 to 1 ratio, you could add about 4 fl. oz. of concentrate (⅓ of a can). That would bring you up to about 28 fl. oz. to 8 fl. oz., a 3½ to 1 ratio which would be better.

To be more exact, you could add 2 more fl. oz. of water to bring the amount of water to 30 fl. oz., and 6 more fl. oz. (½ can) of concentrate to bring the concentrate to 10 fl. oz. This will produce a 3:1 ratio of water to concentrate.

A diagram might look like this:

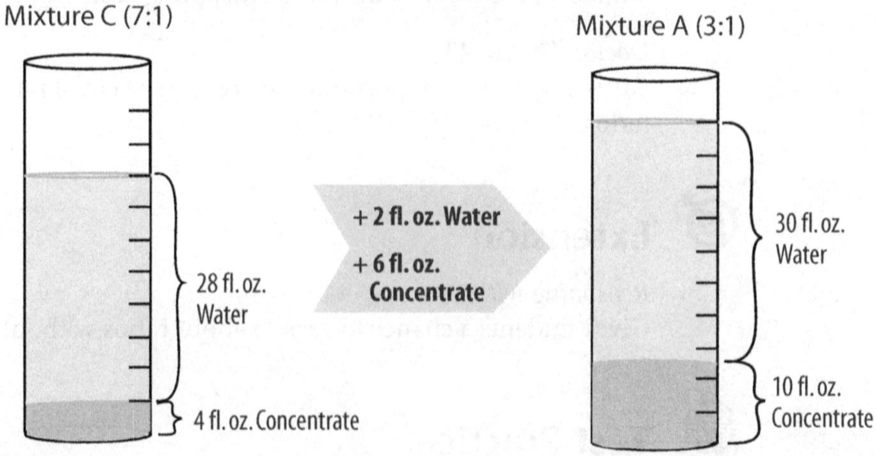

Ask prompting questions to help students think critically about how to reason about this, both with diagrams and numbers. (See *Lesson in Action*, pp. 49–50, for one teacher's facilitation strategy when students got stuck.)

Summary Discussion

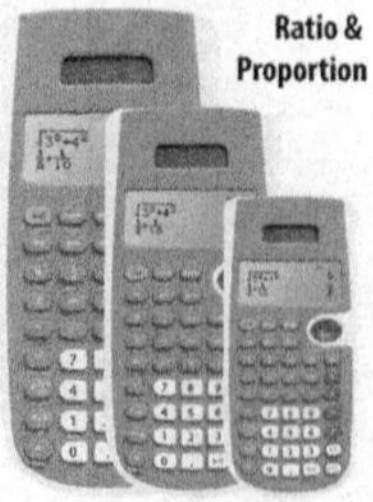

Ask:

How can mathematics help in a situation where a ratio is out of balance?

Students should mention using concrete representations and pictures, using a rule such as doubling both numbers or multiplying both amounts by the same number.

Then ask students to reflect on today's activity in the *Reflections* section (*Student Book*, pp. 141–144).

Heads Up!

Remember to save the mixture displays with students' notes for the next lesson.

Practice

How Does It Taste? p. 38
Provides practice in making predictions about sets of ratios.

Lemonade for the Party, p. 39
Offers more practice with mixtures.

Building a Set of Equal Ratios, p. 40
Gives students more practice with creating equal ratios.

Popcorn Party Mix, p. 41
Students look at the ratio of two ingredients within a recipe.

Dressing It Up, p. 42
Students practice tripling and quadrupling recipes.

Doctor This, p. 43
Gives students an opportunity to create new equal ratios from out-of-balance ratios.

Extension

Reasoning with Ratios, p. 44
Gives students a chance to reason about ratios without using numbers.

Test Practice

Test Practice, p. 46

Looking Closely

Observe whether students are able to

Use ratios to describe taste and visual comparisons

Students should be able to visually distinguish between the mixture that is too weak (too much water) and that which is too strong (too much concentrate). If students don't often drink orange juice, they may have a little more difficulty distinguishing between the mixture with the correct ratio and the weak mixture. The key point is that students realize that ratios are commonly used in everyday life, and when the ratios are out of balance, things are just not "quite right."

Determine how to adjust ingredients so that the proportions are correct

This may be a bit challenging for students, but they can work with what ingredient needs to be added (water or concentrate). Encourage students to draw pictures or actually act out the process in order to figure out how to adjust the ingredients.

Thinking about the ratio of ingredients in a mixture can be especially challenging because one cannot separate the ingredients. To think about the relationship between the amount of water and concentrate, it may be helpful to think about the point at which the mixture was created. Ask students to consider what went into the mixture.

WHAT TO LOOK FOR IN *LESSON 3*	WHO STANDS OUT? (LIST STUDENTS' INITIALS)			NOTES FOR NEXT STEPS
	STRONG	ADEQUATE	NEEDS WORK	
Concept Development • Connects taste and appearance with the idea of ratios • Judges accuracy of predictions based on correct recipe ratios • Given a mixture that is too strong or weak, reasons with ratios to "doctor" it				
Expressive Capacity • Describes the strength of a mixture with ratios • Writes a ratio in several forms (a/b, *a* per *b*, *a* parts of one to *b* parts of the other.) • Can explain clearly how predictions compare to the correct recipe				
Use of Tools • Uses pictures or objects to show that two ratios are or are not equal • Uses a mathematical rule (multiply both amounts by the same number) to compare ratios • Uses a table to keep track of information				
Background Knowledge • Connects personal experience with recipes to the activity at hand				

Rationale

Although everyday experiences are ratio-rich, math textbooks often go straight to the word problems. Here students are asked to connect with the concrete experience and reflect on ratios within that experience. This lesson is rich in mathematics, yet students feel a sense of play as they predict and then determine actual ratios of ingredients.

Math Background

Part-to-Part versus Part-to-Whole

A ratio is a comparison of two things at a time. Because the orange juice activity involves three amounts—water, concentrate, and the resulting juice mixture—students need to stay focused on the two items they are comparing each time. In this lesson, the part-to-part ratio is the central idea.

However, in any situation where a whole is made of parts, usually several ratios are of interest. Consider, for example, a class with an enrollment of 25 students, 10 of whom prefer paper books to e-readers. These are part-to-part ratios. For example, the students preferring books:students preferring e-readers ratio can be expressed as 10 to 15, or 2:3.

Or the e-reader:book ratio can be expressed as 15 to 10, or 3:2.

There are **part-to-whole** (part:whole) **ratios**. For example, the ratio of students who prefer books: total students can be expressed as 10 to 25, 10 out of 25, or 2:5.

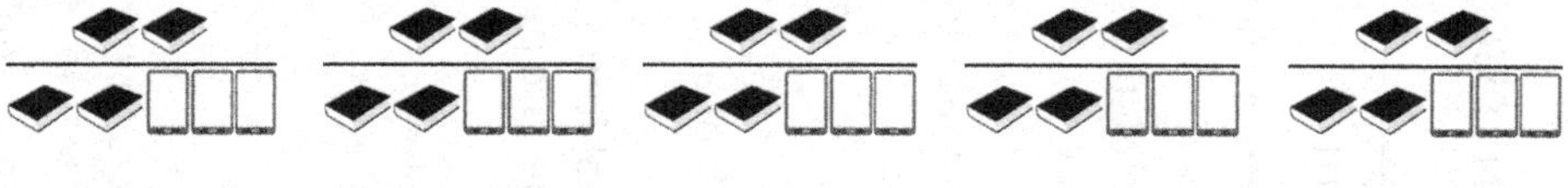

Or the ratio of students who prefer e-readers: total students can be expressed as 15 to 25, or 3:5.

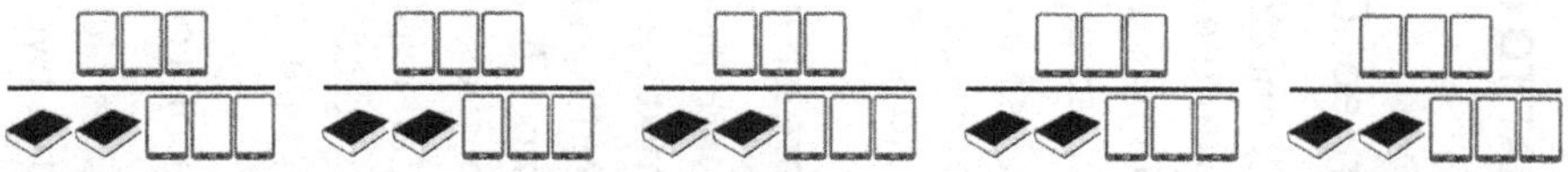

Part/whole ratios are consistent with the common model for fractions and percents, where the numerator is a part and the denominator is the whole (3/5 prefer e-readers and 2/5 prefer books).

This presents some puzzles. First, why do we use 2/3 when comparing part to part and 2/5 when comparing part to whole? This will be further explored in *Lesson 4: Another Way to Say It.*

Secondly, in the part-to-part model, the fractions do not "act" the same as they do in the part-to-whole model. It would make sense to add: 2/5 (the book reader component) + 3/5(the e-reader component) = 5/5 (the whole class). But it does not make sense in this case to add 2/3 (the ratio of students who prefer books to those who prefer e-readers) + 3/2 (the ratio of students who prefer e-readers to those who prefer books).

Using Friendly Fractions

The relative difficulty of the numbers affects how easily ratios can be seen. That is why this unit begins with small whole numbers. However, in this lesson, students are expected to be comfortable handling common fractions such as ½, ¼, ¾, ⅓, and ⅛ (called "benchmarks" in the units *Using Benchmarks* and *Split It Up).* Students are not only asked to consider scaling up by factors of 2, 3, and 10, but are also presented with situations where they need to multiply a recipe by 1½. They might also triple a recipe that starts out with 2½ cups of an amount. Students who have covered the work in the earlier units should be able to do these problems in their heads and support their reasoning with diagrams. It is not necessary to know the arithmetic algorithms for fraction operations.

Context

Recipes are used here as a context for proportional reasoning; proper proportions are everything when it comes to how food tastes and the quality of such things as baked goods. However, written recipes with standard cooking units are not always used. Be sure to acknowledge that many good cooks rarely use written recipes and that recipes may be more common for baked goods. Also note that it is sometimes difficult to imagine the ratios in mixtures such as orange juice because everything is mixed together. The ratios cannot be empirically proved because one cannot physically separate the concentrate from the water once everything has been mixed together.

Facilitation

Allow students plenty of time to explore and make predictions. Students often begin to gain a more positive attitude about math when they see how it relates to their own lives and how it can be fun. Connecting math to the sense of taste can be a rewarding experience if they are allowed to explore.

One student commented at the end of this lesson, "I feel much more comfortable tackling math problems. I'm having fun, and now I see ratios everywhere. I really like these lessons—the fact that we're focusing on different senses."

Making the Lesson Easier

If students have predicted the amounts using fractions (such as a ½ cup of water), encourage them to change the amounts to fluid ounces. This will allow them to work with whole numbers rather than fractional parts.

Because it is difficult to "see" the individual ingredients in the orange juice mixture, you might want to begin with a different recipe. A trail mix, fruit salad, or bean salad is an alternative way to help students see the implication for changes in ratios.

Making the Lesson Harder

Challenge students to think proportionately without numbers—in other words, qualitatively rather than quantitatively. Assign *Extension: Reasoning with Ratios* (*Student Book,* p. 44).

Alternatively, ask students to compare the two recipes (*Blackline Masters 4* and *5*) and come up with yet another way to arrive at the three mixtures in the ratios of 3:1, 1:1, and 7:1. There is more than one way to produce juice mixtures of these strengths.

These are the predictions from one group of taste testers when we piloted this activity.

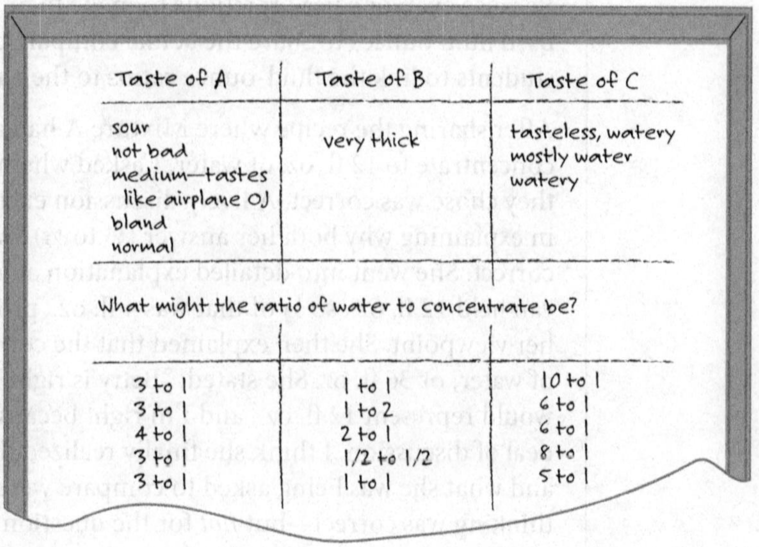

Mary Jane Schmitt
TERC, Cambridge, Massachusetts

Students entered into the discussion with gusto; however, keeping track of the three amounts—the concentrate, the water, and the mixture—was sometimes a challenge.

Three students were in class that day, so I only used one large can of orange juice concentrate. This got me, the teacher, thinking proportionately right from the start while preparing to teach this lesson. I divided the concentrate into three equal parts, which gave me 4 fl. oz. per mixture.

Recipe A	Recipe B	Recipe C
Concentrate: 4 fl. oz.	Concentrate: 4 fl. oz.	Concentrate: 4 fl. oz.
Water: 12 fl. oz.	Water: 4 fl. oz.	Water: 28 fl. oz.

After everyone taste-tested, we captured the predictions on chart paper. The students all used ⅓ of the can of concentrate as a basis for comparison because they knew that I had used only one can and created three mixtures.

Mixture	Betty	Laura	Terri
A	1/3 can concentrate to 1 can water	1/3 can concentrate to 2/3 can water	1/3 can concentrate to 1/3 can water
B	1/3 can concentrate to 1/2 can water	1/3 can concentrate to 1/3 can water	1/3 can concentrate to 1/12 can water
C	1/3 can concentrate to 11/2 can water	1/3 can concentrate to 1 (3/3) can water	1/3 can concentrate to 2/3 can water

continued on next page

continued from previous page

Because everyone used fractions to make their predictions, I intentionally used fluid ounces to share the actual comparisons. I wanted to challenge the students to link the fluid-ounce ratios to their ratios as fractions.

After sharing the recipe where Mixture A had a ratio of 4 fl. oz. of concentrate to 12 fl. oz. of water, I asked whether either of the predictions they chose was correct. A lively discussion ensued, with Terri quite articulate in explaining why both her answer (⅓ to ⅓) and Betty's (⅓ to 1) were correct. She went into detailed explanation about the fact that the entire can held 12 fl. oz., so ⅓ of that was 4 fl. oz., providing a drawing to support her viewpoint. She then explained that the correct mixture used three cans of water, or 36 fl. oz. She stated, "Betty is right because one can of water would represent 12 fl. oz., and I'm right because ⅓ of 36 is 12." After a great deal of discussion, I think she finally realized that what she was comparing and what she was being asked to compare were two different things. Her thinking was correct—but *not* for the question posed.

Everyone agreed that Mixture B was pretty thick, that I had probably added too much concentrate or too little water. I showed them the ratio of 4 fl. oz. concentrate/4 fl. oz. water and then asked them whether either of the predictions was correct. Although Laura's answer was right, Terri had used the same logic she had used earlier, so her reasoning was actually quite sound (although incorrect, given the items being compared). When she was working on her prediction, she had asked me what half of ⅓ was. I asked her what ½ of a ½ was, then ½ of a ¼. In each instance, she was able to give me the correct response. I then asked what ½ of ⅓ was. She tentatively replied, "One-sixth." Once I let her know that she was right, she figured out that ½ of ⅙ was 1⁄12. (One-twelfth of 36 fl. oz. is 3 fl. oz., which is what she had reasoned was the amount of water in Mixture B.)

For Mixture C, everyone agreed that the mixture was very watery, but no one was able to accurately predict just how much water there was. I put on the board the ratio "4 oz. concentrate/28 oz. water." Terri explained her thinking again (⅔ to her represented ⅔ of 36, which would be 24 fl. oz.—close). We then talked about how much of the total amount of water needed 28 fl. oz. represented. Terri drew a picture:

xxxxxxxxxxxx / xxxxxxxxxxxx / xxxxxxxxxxxx

She then said, "28 is two whole cans of water (12 fl. oz. each) and four more of the 12-oz. can, which reduces to ⅓."

Donna Curry
Gardiner Adult Education Program, Gardiner, ME

FACILITATING LESSON **4**

Another Way to Say It

How are ratios used to influence others?

Synopsis

The language of ratios is the central theme of this lesson. Students increase their repertoire of ways to describe comparisons, particularly multiplicative ones. In doing so, they focus on part-to-part as well as part-to-whole relationships. Percents are introduced as a way to indicate a part-to-whole relationship.

1. The class continues from the last lesson considering the different ways that they might write comparisons between amounts of frozen concentrate, water, and orange juice mixture.
2. Students then write "two truths and a lie" about the orange juice mixture, and student pairs try to figure out which is which.
3. Finally, students apply the different ways of writing comparisons in advertisements.

Heads Up!

This lesson assumes students are comfortable with common percents. If students have not had previous exposure to benchmark fractions and percents, you may want to introduce them to benchmark percents before beginning this lesson. See the EMPower units: *Using Benchmarks: Fractions and Operations* and *Split It Up: More Fractions, Decimals, and Percents.*

For additional practice outside of class, consider sharing with students the ratios portions of this free, downloadable packet Being Counted: Probability & Statistics. https://www.collectedny.org/wp-content/uploads/2020/02/Being-Counted-Probability-Statistics-Fast-Track-GRASP-Math-Packet-Part-1-v1.6-10.04.2021.pdf

Objectives

- Understand alternate but equivalent ways to make a comparison of two amounts
- Use percents as ratios to compare the part to the whole amount
- Round whole numbers to make comparisons more manageable

Materials/Prep

- The display students contributed to from the previous lesson on orange juice comparisons

For *Activity 3*, prepare to display *Making Advertising Statements* (*Blackline Master 6*) to refer to during the discussion.

Opening Discussion

Review the last lesson in which students taste-tested and compared orange juice mixtures. Display their findings to use as a starting point for discussion.

Remind students that the correct ratio of the amount of water to the amount of concentrate was 3:1. Then ask:

What percent of the mixture was concentrate? What percent was water?

How would you show it with a picture?

Where do you see the "3" and the "1" (from the ratio of 3 to 1) in your picture? Where do you see the percent?

Heads Up!

Anticipate that some students will come up with 33 1/3% and some will come up with 25% as an answer. This dilemma is the core of this lesson. Insist on a visual representation so that all students can see that the ratio of 3 parts water to 1 part concentrate in this example actually yields a mixture that is 25% concentrate and 75% water.

Activity 1: Part-to-Part versus Part-to-Whole

Refer students to *Activity 1*, p. 48, in the *Student Book*. Ask them to consider how many different ways the three items—water, concentrate, and juice—can be compared, two at a time. Ask them to write out each comparison in words and numbers and with pictures.

They might make observations such as these:

- amount of concentrate to amount of water (1:3)
- amount of concentrate to amount of juice (1:4)
- amount of water to amount of concentrate (3:1)
- amount of water to the amount of juice (3:4)

If students have trouble coming up with comparisons between part and whole (such as amount of concentrate to amount of mixture), suggest they draw a bar model such as this:

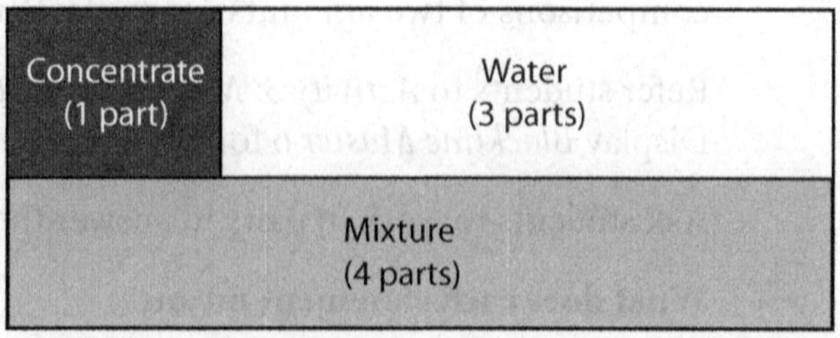

Students might also make additive comparisons rather than multiplicative ones, for example:

> "There are two more cans of water than concentrate."
>
> "In Mixture C, there are six more cans of water than concentrate."

Although these comparisons are true, they are not ways to state ratios. Remind students that although these statements are valid, they are not examples of comparisons using ratios.

Activity 2: Two Truths and a Lie

Refer students to *Activity 2* (*Student Book*, p. 49). Direct them to think now about the failed mixtures and the ratios of ingredients used to create Mixture B and Mixture C. Ask students to use ratios to write two true statements and one false statement for Mixture B and/or Mixture C. Encourage them to use a percent in at least one of their statements.

Then ask students to pair up and exchange their statements. Each student will try to figure out which are his or her partner's truths and which is the lie. Suggest that students use pictures to illustrate why some statements are true and some are false.

By now students should be able to write true statements such as the following:

> "The ratio of the amount of concentrate to the amount of Mixture C is 1:7."
>
> "The percent of concentrate in Mixture C is 12 1/2% (or 12.5%) because the ratio of concentrate to mixture is 1:8."
>
> "Fifty percent of Mixture B is concentrate."
>
> "The amount of concentrate to the amount of water in mixture B is 1:1."

Percents are based on part to whole relationships. Watch students as they create percents. If they compare part to part (concentrate/water), ask them to think about the idea of percents and what that idea represents: thinking of the mixture as the whole (100) and then comparing the part to 100.

Activity 3: Making Advertising Statements

Ask students to think of some advertisements they have heard or seen in which comparisons of two amounts are made. Write down their ideas on the board.

Refer students to *Activity 3: Making Advertising Statements* (*Student Book*, p. 50). Display *Blackline Master 6* for all to see.

Ask students to work in pairs to answer the following questions:

What does each statement mean?

How can you show what each means using a picture?

What are the relationships, if any, among the four statements?

Which of the four statements do you think would be most useful as an advertisement for Harvest Blend Bread? Why?

When pairs are done, discuss the results as a class. They should share their reasoning and their drawings.

The four ads convey the same information, but this is not at all obvious. Drawings and the property of equal fractions should be employed as tools to support reasoning.

Summary Discussion

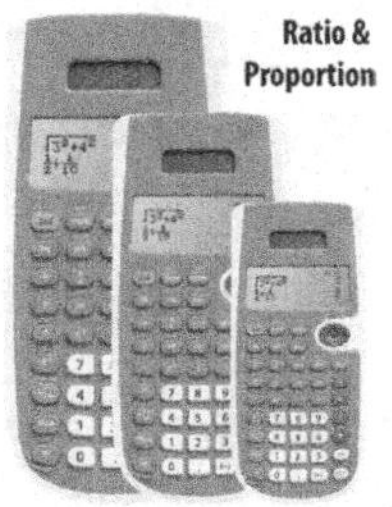

Ask students to share some of the ways they can talk about comparisons of amounts. Be sure that they mention the idea of percent as ratio (between part-to-whole rather than part-to-part).

Then refer students to the *Reflections* section of the *Student Book*, pp. 141–144.

Practice

What Is Another Way to Say It? p. 51
Gives students practice creating ads with similar claims.

Keeping the Ratio the Same with Gardeners, p. 52
Provides practice writing similar ratios.

Who Is Right? p. 53
Provides practice comparing part:part to part:whole relationships.

Rounding Ratios, p. 54
Practice for those students who need to review rounding.

Test Practice

Test Practice, p. 55

Looking Closely

Observe whether students are able to

Understand alternate but equivalent ways to make a comparison of two amounts

Are students aware of the difference between multiplicative and additive comparisons? They should be able to compare using both additive methods (looking at the difference between two amounts) and multiplicative methods (looking at the ratio of one amount to another). Help students understand that both methods are useful, but only one represents comparisons using ratios.

Although students have probably often heard different ways to make comparisons, they may have limited exposure to alternate but equivalent ways to say the same thing. They may have difficulty differentiating between part/part relationships and part/whole relationships. For example, three parts water out of a total of four parts mixture means that 75% of the mixture is water. Yet, there are three parts water to one part concentrate for a 3 to 1 ratio.

Visual representations will help ground comparisons. For example, the number of people who prefer Harvest Blend to those preferring Sunshine Sliced Bread is 4:1. A visual would show clearly the total number of people:

HHHHS = total of 5 people

or

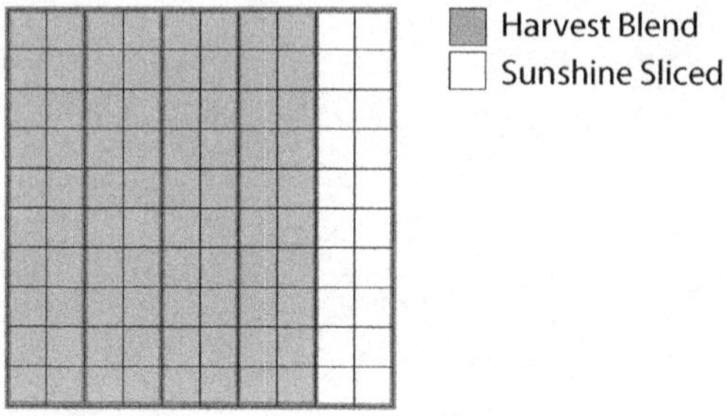

A simple visual shows the comparison of two items, while at the same time showing the total number of items that make up the two sets of items. For example, a visual for the statement “80% of those polled prefer Harvest Blend to Sunshine” helps students see the whole picture as well as the relationship between the two items:

HHHHHHHHSS = 8 out of 10 or 80%

$$\frac{\text{Prefer Harvest Blend}}{\text{Poll results}} = \frac{\text{HHHHHHHH}}{\text{HHHHHHHHSS}} = \frac{\text{HHHH}}{\text{HHHHS}} \quad \frac{\text{HHHH}}{\text{HHHHS}} = \frac{8}{10} = \frac{4}{5} = \frac{80}{100}$$

A circle graph could also show equivalence:

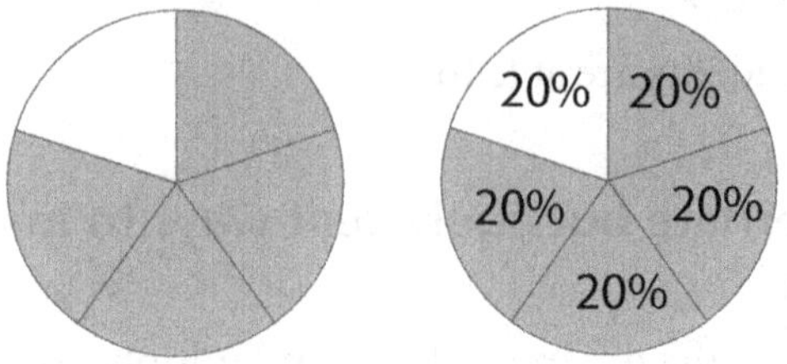

Use percents as ratios to compare the part to the whole amount

By the time students get to this unit, they should have had multiple experiences with benchmark percents. They should be able to move quickly from fractions representing half to 50% and should be able to estimate other benchmark percents based on 10%. Calculators could be used to check work.

Even though this lesson includes some unwieldy numbers, the expectation is that students will round them in order to write ads with more friendly-sounding numbers.

Round whole numbers to make comparisons more manageable

The whole numbers in this lesson get a bit more complex, but, in reality, polling reports tend to round numbers when they report them. Students are encouraged to regard situations such as "5,975 people preferred Harvest Blend Bread and 1,496 people preferred Sunshine Sliced Bread" as the approximate equivalent of 6,000 to 1,500.

WHAT TO LOOK FOR IN *LESSON 4*	WHO STANDS OUT? (LIST STUDENTS' INITIALS)			NOTES FOR NEXT STEPS
	STRONG	ADEQUATE	NEEDS WORK	
Concept Development • Understands the part:part relationship and the part:whole relationship in a mixture or statistics from a survey • Given a part:part relationship, can see the part:whole relationship • Given a part:whole relationship, can see the part:part relationship				
Expressive Capacity • Accurately describes the strength of a mixture in various ways with ratios and percents • Accurately describes the relative relationships of categories of survey data • Writes a part:whole ratio in several forms (a/b, $a{:}b$, x percent) • Can explain clearly why different representations are or are not equivalent				
Use of Tools • Uses pictures or objects to show part:part and part:whole relationships • Uses pictures or a mathematical rule to show the equivalence of a ratio in fraction form and a percent				
Background Knowledge • Is aware of the ways advertisers use survey data to persuade consumers				

Rationale

Students need to be flexible in their thinking about ratios. Seeing or hearing ratios only as fractions or as unit rates limits their understanding of the world around them. By providing opportunities to practice making comparisons between two parts and between one part and the total, students become more critical thinkers as they learn to listen carefully and question what they see and hear.

Math Background

Percents and rounding are the two number-related topics introduced in this lesson. The percent ratios are kept simple here. Messier percents will be addressed later in this unit. This lesson focuses on the use of percent as another way to talk about ratios in which one item is compared to the total amount.

Rounding results in whole numbers. Students need to think about how to round each amount in order to make manageable comparisons. Therefore it will not always make sense to round each amount to the nearest whole hundred or thousand; it might be more manageable to round one of the amounts to the nearest thousand and the other to the nearest hundred. Help students see rounding as flexible (not a rule to follow) and done by thinking about what would make the problem easier, given the comparisons they are making.

Context

The language and the psychology of statistical reports go hand in hand. Slowing down to think about what a report is actually saying is a skill required of any numerate adult. As students work through the language of this lesson, ask them to bring in articles or advertisements that use ratios to make claims. Discuss not only the number relationships conveyed, but also the possible intent of the writer.

Facilitation

If students are having difficulty, encourage them to make the numbers and words come alive by making a drawing. Or use the class as an example and create various ratios using part:part situations as well as part:whole class comparisons. At first, allow students actually to line up to create groups to compare, and then jot down the relationships numerically on the board. For example, a part:part relationship could be the number of students wearing glasses vs. those who are not. Part:whole relationships could include the number of glasses-wearers vs. the total number of students.

Making the Lesson Easier

Use concrete objects to make simple comparisons. Build from the visual representations, but also make a connection to the property of equal fractions.

Making the Lesson Harder

Have students create and conduct their own survey. Then have them report their findings using a variety of comparisons including part:part and part:whole.

Students struggle with the differences between the part/part and part/whole concept. One teacher wrote:

> Visual representations saved the day for all of us! Students struggled with this lesson because they weren't totally fluent with their number facts and mental math. We worked on the idea of part:part vs. part:whole for a couple of days. This is an example of how the visual representations helped students as they created new ways to express ratios:
>
> 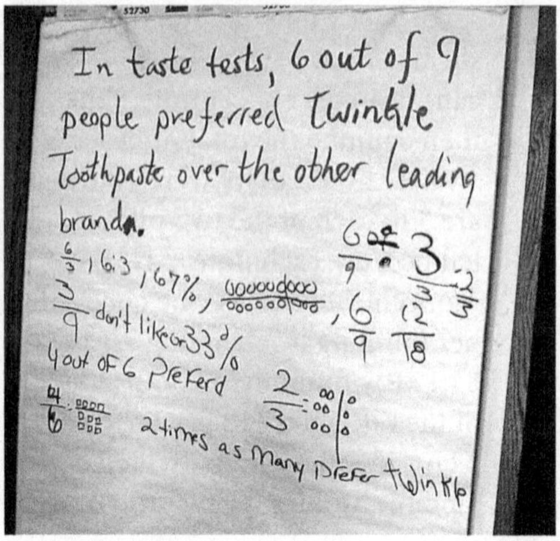
>

One student grappled with the idea of Mixture C having a ratio of 7 parts water to 1 part concentrate by using drawings to convince herself.

> Lynette became stuck on Mixture C. In her *Student Book*, the ratio was represented as 7 cans water to 1 can concentrate. She was clear that was not the same ratio as 28 oz. water to 4 oz. concentrate. I asked her to prove me wrong to help me understand how I had incorrectly mixed Mixture C.
>
> She tackled the problem by considering what the ratio would have been if I had used only 1/3 of the total can of concentrate instead of a whole can. She figured out that the ratio of 7 cans of water to 1 can of concentrate was the same as 2 1/3 cans of water to 1/3 can of concentrate by visually cutting her cans into thirds.
>
> *continued on next page*

continued from previous page

Next she had to figure out whether the ratio of water to concentrate—2 1/3 to 1/3—was the same as 28 to 4. Because she was stumped but was very attuned to visual representations, I suggested she use sticky notes to figure it out. She marked some notes to represent 1/3 part concentrate. She then decided to represent each of her whole portions of water as three 1/3 sticky notes. She discovered that she "proved" that 7:1 was the same as 2 1/3 to 1/3 and 28 to 4. (The following set of sticky notes shows that she has 7 parts water [each = 1/3 can] to 1 part concentrate.) She then counted the total number of "water" notes; there were 28 (there are 3 in each of the two piles on the left because each note = 1/3) to 4 "concentrate" notes. She checked her thinking by changing 7:1 to 84:12 (by multiplying both sides of the ratio by 12 to get a total of 12 oz. in a can), then divided each amount by 3 for a new ratio of 28:4.

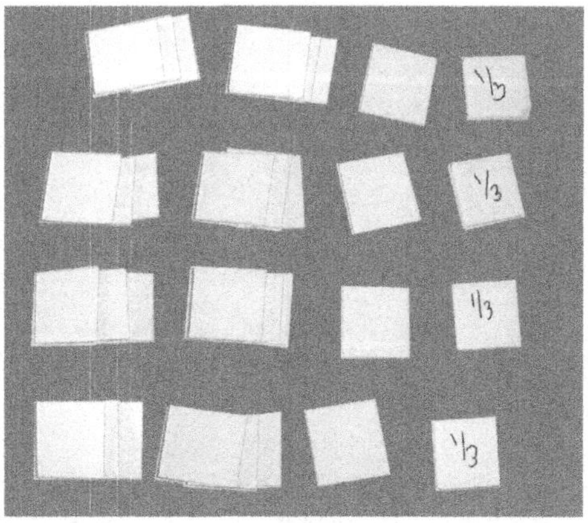

Yet another student tackled the same dilemma in a different way. She considered the "right way" to make the orange juice mixture and drew it using 4-oz. representations. She then used the same idea to create Mixture C using 4-oz. representations (notice top section of the following picture). She had 12 oz. of concentrate but then erased two of the three sets of 4 oz., which gave her 4 oz. remaining on the left. On the right, she created three sets of 4 oz. to make one set of 12 oz. She did this twice and then added another 4 oz. This gave her 28 oz. of water, or seven sets of 4 oz.

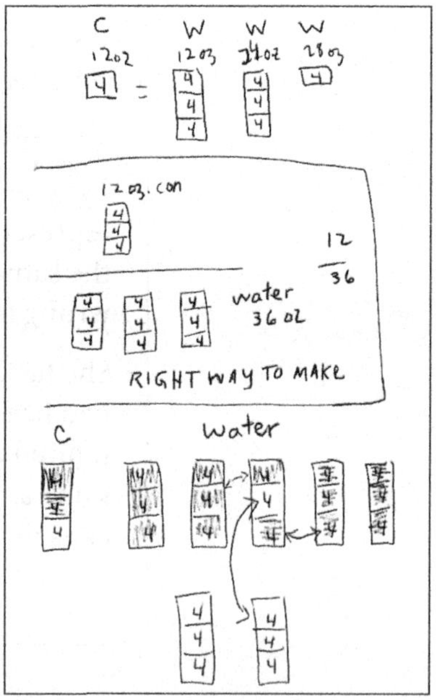

She then double-checked her answer by drawing another visual. Notice the colors match; she has 2 1/3 parts of water to each 1 part concentrate.

I am fascinated by how easy it is to visualize with either of their representations that 2 1/3 to 1/3 is the same as 7 to 1, which is the same as 28 to 4.

Donna Curry
Gardiner Adult Education Program, Gardiner, ME

FACILITATING LESSON 5

Mona Lisa, Is That You?

Do the proportions appear correct?

Synopsis

The focus of this lesson is on proportionality as it applies to stretching and shrinking two-dimensional pictures. Students first use their visual sense to consider picture proportions and then verify the proportionality with numbers. **Graphing** is introduced as another tool for exploring and confirming equal ratios.

1. Given a 2″ x 3″ miniature reproduction of the *Mona Lisa* and the estimated actual length of 30″, students predict the actual width.
2. Students visually discriminate between good reproductions and distorted ones, then test their perceptions with measurements.
3. In pairs or small groups, students explore the use of a graph as one way to test proportionality.
4. The class comes together to connect the numerical dimensions with the points on a graph and the three regions where the points fall: the good ones on the line, the skinny ones on one side of the line, and the too wide ones on the other side of the line.

Objectives

- Judge by eye whether variously sized reproductions are proportional to an original
- Determine whether reproductions are actually proportional, using measurement
- Use a graph as a tool to test for good reproductions (equal ratios)
- Connect the graph points with the number pairs

Materials/Prep

- Calculators
- Colored markers
- Graph paper
- Easel paper (preferably with grids)—one sheet per pair
- Scissors
- 12″ rulers marked in cm and inches
- Yardsticks

For each small group, prepare one set of *Mona Lisa Cards* (*Blackline Masters 7–11*). Ask students to cut out the seven pictures of *Mona Lisa* before working with them, marking each with its letter. To save time, you might want to cut out the pictures in advance.

Heads Up!

This lesson has several parts. Plan for more than one session.

Opening Discussion

Introduce the lesson by saying:

In the orange juice activity, you used your sense of taste to judge whether ingredients were in the right or wrong proportions. Today you will use your visual sense.

Activity 1: How Large Is the *Mona Lisa* Painting?

Start by asking students to examine the small image of *Mona Lisa* in *Activity 1* (*Student Book*, p. 58). Say:

Here is a fairly good miniature reproduction of the *Mona Lisa*. If the original *Mona Lisa* measures 30 inches long, how wide would you estimate the real painting to be?

Post all answers, and ask students to draw the size of the original *Mona Lisa* on a piece of large easel paper (one piece of easel paper for each group). Be sure to keep these 20″ × 30″ drawings, as they will be used later for the graphing activity. Tell students to verify their answers with diagrams.

Ask:

Where do you see sets of ratios in this activity?

How does the ratio of the width to the length in the original *Mona Lisa* painting compare to the ratio of the width to the length in the miniature reproduction?

How do you know?

Some students might do one of the following:

- Say that "you just multiply 2 by 10 to get 20." If so, ask these students to show with a diagram why that makes sense.
- Show that because there are 10 3-inch segments that fit into the length, 10 2-inch segments might fit into the short length, making the width 20 inches.

If students arrive at an answer close to, but not exactly, 20 inches, ask them to measure off how many 2-inch sides will be needed to measure 20 inches and to try the same number of 3-inch pieces for the length.

The easel paper will look as follows:

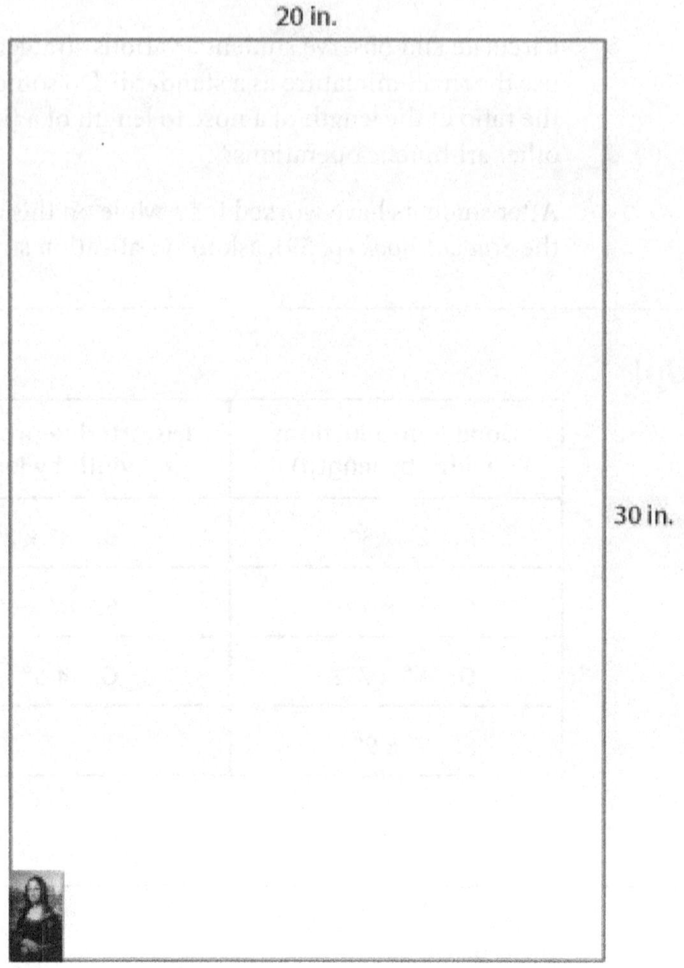

Heads Up!

The actual dimensions of the *Mona Lisa* are 53 cm by 77 cm, or 20.8" x 30". For this exercise, we are calling the dimensions approximately 20" x 30".

Activity 2: Mona Lisa, Is That You?

Part 1

In addition to the 2″ × 3″ "good" *Mona Lisa* and the full-size outline, furnish each group with a set of six reproductions (for a total of seven). Working in small groups, students need to agree on two piles: good reproductions and distorted reproductions. They should do this by "eyeballing" the pictures. They might want to make a third pile for pictures they are unsure about.

After groups agree on the piles, have them verify their decisions. Ask:

How can you use numbers to test whether there is distortion?

Circulate and observe students' various strategies. Do some use rulers? Or do they use the small miniature as a standard? Do some people compare features such as the ratio of the length of a nose to length of a finger? Or do they use doubling or other arithmetic operations?

After students have worked for a while on this and have recorded their work in the *Student Book* (p. 59), ask for verification strategies.

Heads Up!

Good Reproductions (width by length)	Distorted Reproductions (width by length)
A. 2″ x 3″	B. 3″ x 2″
C. 3″ x 4½″	E. 6″ x 6″
D. 5″ x 7½″	G. 4.5″ x 9″
F. 6″ x 9″	

Part 2

Ask students to create a set of equal ratios with the dimensions of their good reproductions. They should be prepared to show that each ratio is the same as the original mini-reproduction. For example:

$$\frac{\text{width (in inches)}}{\text{height (in inches)}} \quad \frac{2}{3} = \frac{5}{7\frac{1}{2}} = \frac{6}{9} = \frac{3}{4\frac{1}{2}}$$

Ask:

What patterns do you see in these numbers?

Where can you see the 2 to 3 ratio in the good reproductions?

Some people might say, "For every two chunks in the width, there have to be three of those same chunks in the length."

$$\frac{2}{3} = \frac{5}{7\frac{1}{2}}$$

("I see 2 chunks of 2½ in.")

("I see 3 chunks of 2½ in.")

Some people might say, "For every chunk of 2 inches in the width, there needs to be a 3-inch chunk in the length."

$$\frac{2}{3} = \frac{5}{7\frac{1}{2}}$$

("I see 2½ chunks of 2 in.")

("I see 2½ chunks of 3 in.")

Some people might relate the ratio to the general rule (property of equal fractions) that you can multiply or divide both amounts by the same number and keep the same ratio/balance/relationship.

Ask for a picture to verify this. If students have trouble figuring out how to draw the relationships, you might share this strategy as an example:

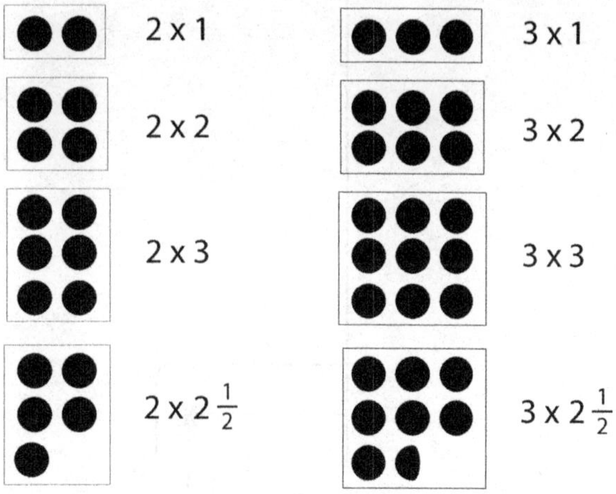

What does it mean to multiply something by 2½? If this is the first exposure that students have to multiplying by a non-whole number, you may want to provide further practice. In this lesson, only mixed numbers with benchmark fractions (such as 1½ and 2½) are used. If students struggle, encourage them to draw visual representations or to think about how they would multiply the amounts if they were units of money. Most students are comfortable multiplying $2.50 by 5 to get $12.50, for example. Ask them to consider what strategies they used to figure out such calculations and advise them to use similar strategies when using fractions rather than dollars and cents. What would it mean to earn $5.00 2½ times?

Activity 3: Lining Up the Reproductions

Part 1

Now make the connection between proportionality and the linear graph. Ask groups to line up the lower left-hand corners of all the good reproductions, including the original, and lay them on the 20” × 30” paper. Tell them to trace the outline of each shape on the easel paper, keeping the bottom left corners lined up at the origin.

Say:

Mark the upper right-hand corner with a star.

What pattern do you notice?

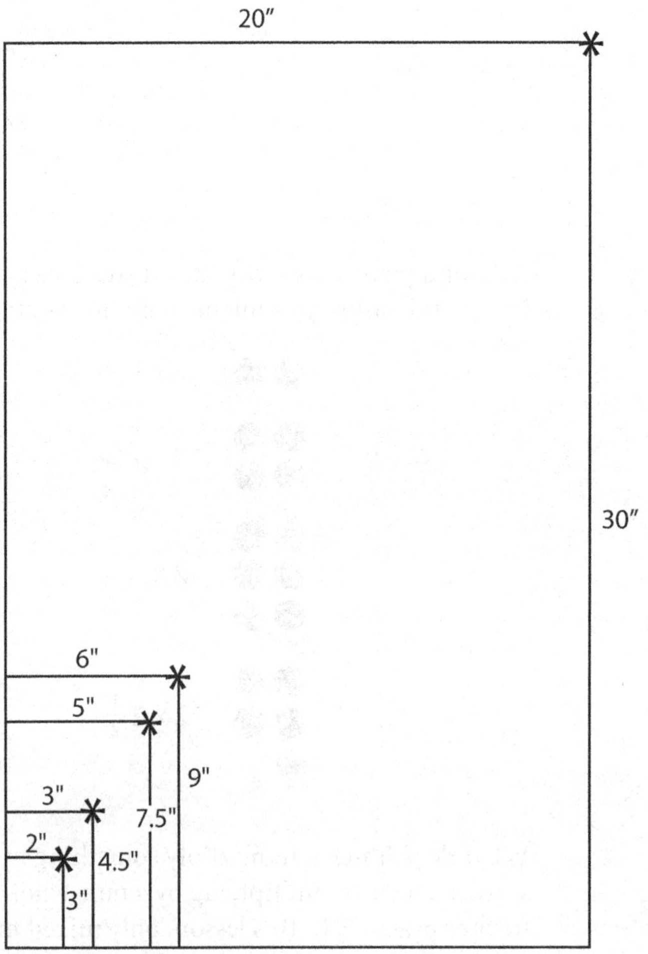

Next consider the distorted pictures.

Say:

With a different color, mark the upper right-hand corner with an "X."

What do you notice?

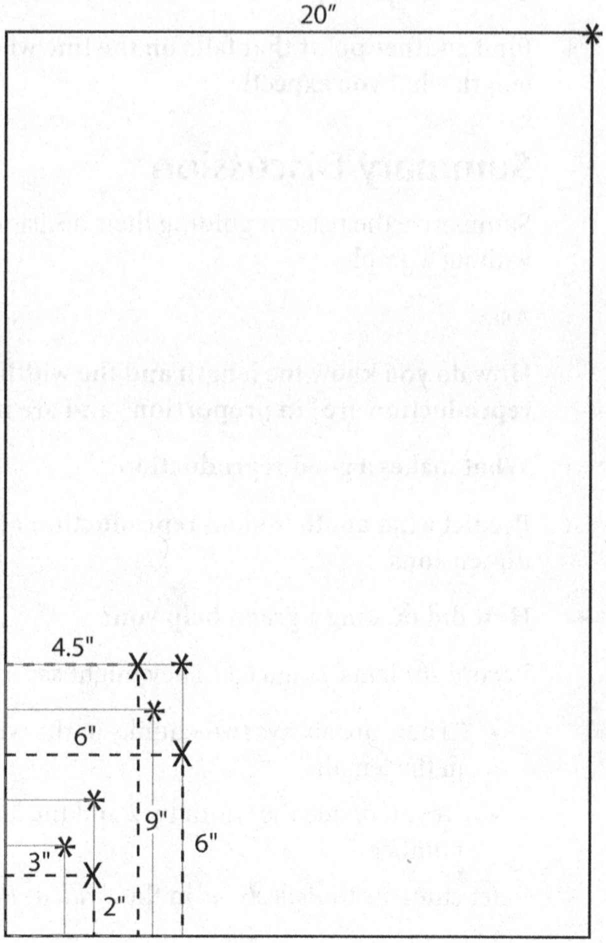

Part 2

Guide students to transfer this information to a graph in *Activity 3* (*Student Book*, p. 62), labeling axes—height by width—and plotting points.

Heads Up!

If students are not familiar with coordinate graphing, have the class do this step together. Introduce the convention of labeling (*x*, *y*).

Summarize:

What does it mean if the corner point falls on the line?

What do the points that fall beneath the line tell you? What about those above the line? Show that these are not in the ratio of 2:3.

Use the graph to predict the dimensions of another good reproduction.

Find another point that falls on the line where the width is 8 inches. Is the length what you expect?

Summary Discussion

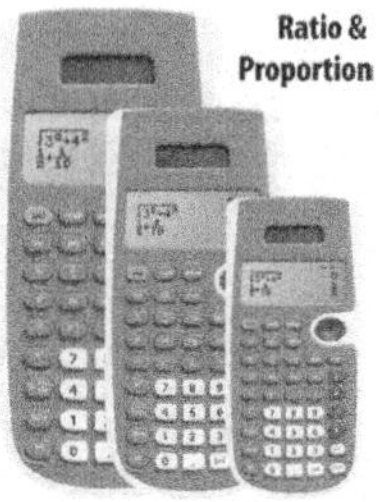

Summarize the reasons guiding their decisions, both with and without a graph.

Ask:

How do you know the length and the width of a reproduction are "in proportion" and are in the ratio of 2:3?

What makes a good reproduction?

Predict what another good reproduction of the *Mona Lisa* might have as dimensions.

How did making a graph help you?

Record students' language. They might say something such as,

- "There are always two chunks in the width to three of the same size chunks in the length."
- "If you divide the width by 2 and the length by 3, you will get the same number."

Refer students to *Reflections* in the *Student Book* (pp. 141–144).

Practice

Rectangle Ratios, p. 63
Provides practice in figuring out whether ratios are equal.

What Is the Pattern? p. 64
Gives further practice looking at patterns for those students who need it.

Using a Graph to Predict, p. 65
Gives practice in using graphs to predict for another situation.

Graphing Ingredients, p. 66
Gives practice looking at graphs to determine whether ratios are in proportion.

Extension

Extension: Paint by Numbers, p. 58
Students look at a set of six differently sized *Mona Lisa* reproductions and determine which are good and which are distorted.

Test Practice

Test Practice, p. 70

Looking Closely

Observe whether students are able to

Judge by eye whether variously sized reproductions are proportional to an original

Students should be able to visually compare two pictures to determine whether they are fairly proportional. If they struggle, you may want to try creating some distorted vs. good reproductions using a word-processing program. Students should be able to articulate why the pictures look distorted. Are they long and narrow or are they wide and squat?

Determine whether reproductions are actually proportional, using measurement

Students have been using pictures and equal ratio patterns to determine whether ratios are proportional. In this lesson, they acquire a new tool but continue to use their other strategies to confirm whether reproductions are proportional.

Use a graph as a tool to test for good reproductions (equal ratios)

See whether students are able to consistently place all of the good reproductions along the same corner and whether they face each picture in the same direction. This is a visual example of how order matters. If students place a 2″ × 3″ reproduction lengthwise at a corner and then a 4″ × 6″ reproduction widthwise, they will not be able to draw a straight line connecting end points from (2, 3) through (4, 6) and (20, 30). If students have not been consistent, give them time to discover for themselves what happened.

Connect the graph points with the number pairs

Can students describe the corner points of each of the pictures? If students are inexperienced in creating and reading straight-line graphs, take time to explore other points on the graph. Ask them to find other points using just the coordinates; then have them find a point on the graph and call out the coordinates. It might also be helpful to relate the ability to read coordinates to map reading, a real-life skill.

WHAT TO LOOK FOR IN *LESSON 5*	WHO STANDS OUT? (LIST STUDENTS' INITIALS)			NOTES FOR NEXT STEPS
	STRONG	ADEQUATE	NEEDS WORK	
Concept Development • Reasons proportionately to judge whether a picture reproduction is good or bad • Given one picture, can project to a picture with proportional measurements • Attends to the order within the ratios (e.g., 2"/3" vs. 3"/2")				
Expressive Capacity • Accurately describes the relative relationship of the width to the length of a picture • Can explain clearly why the various pictures are or are not proportionate				
Use of Tools • Accurately measures with a ruler • Uses number relationships to verify proportionality • Creates and uses a graph to explore proportionality				
Background Knowedge • Draws on past experience with distorted visual images				

Rationale

This lesson is yet another opportunity to explore proportionality in a hands-on way. The more comfortable students are with "playing" with proportional reasoning using a variety of senses, the more comfortable they will be handling proportion problems in less familiar contexts later.

The dimensions of length and width lend themselves to a graphed representation of increasing sizes, but the relationship of any two elements of a ratio can be transferred from a fraction to x- and y-axes. Once a pair of equivalent ratios is plotted and the points connected with a line, equivalent and nonequivalent ratios are visually obvious without a lot of calculation. Points that fall on the line are in proportion; those that do not land on the line are not in proportion.

Math Background

Two significant mathematical ideas are at the core of the *Mona Lisa* lesson: similarity and coordinate graphing. Both ideas recur throughout more formal mathematics courses, and this lesson is designed to offer an experience to help ground those concepts.

Similarity

Similarity is a topic where geometry and proportion come together. Similar polygonal shapes, whether they are rectangles, triangles, or hexagons, have the same shape, but not the same size. An intuitive sense of similarity is called upon when students sort the various *Mona Lisa* pictures. When students begin to produce the ratios of the lengths to the widths, their attention is drawn to the relationship *within* the ratios: *Mona Lisa*'s length is always 1½ times the width. Also, they attend to the relationship *between* the ratios: Doubling the width necessitates a doubling of the length ($2/3 \times 2/2 = 4/6$).

Coordinate graphing

This lesson treats the graph as a tool with which to record and judge proportionality. It is not a formal introduction into coordinate graphing of functions, which students explore in more depth in the *EMPower* unit *Seeking Patterns, Building Rules: Algebraic Thinking.*

As you work with this lesson, you might want to make some connections to the basic algebra students may have learned previously in school.

The graph for any proportional situation is a straight line. Not only is the line straight, but it also slants upwards toward the right and passes through the origin. This is true for graphs of all direct proportional situations. Therefore, straight-line graphs that begin at (0, 0) and have positive slopes can be used as a tool for judging whether a set of ratios is equivalent or not.

Think of the general equation for a straight-line function: $y = mx + b$. If the y-intercept (b) is zero, the equation can be written $y = mx$. The constant m represents the constant multiplicative relationship between y and x. Thus, m (or m/1) can be looked at as the constant of proportionality: $m = y/x$.

Context

Visual contexts provide many opportunities to explore proportionality. Scale drawings used in design (the main focus in *Lesson 6*) depend upon keeping the ratios of all aspects of the design equal. It is the central idea behind maps, graphic design, interior decoration, and architecture.

There are examples in literature, film, and television of protagonists who are giants or are very small relative to other people in the story. Ask students, "Have you ever read a book or seen a movie where there were giants or people who were very small?"

Traditionally, in math text books, problems dealing with trees casting shadows, boats crossing rivers, and the rise and run of stairs are often the visual contexts for proportional reasoning.

Facilitation

There are many ways to detect proportionality, so it is important that you allow students to explore. Encourage them to talk about their strategies, share them with their fellow students, and challenge themselves to look at other strategies for looking at the proportions of the *Mona Lisa.*

Making the Lesson Easier

If students struggle with numbers such as 5 x 7½ or 3 x 4½, leave these reproductions out for now. Have students work only on reproductions that involve whole number multiples.

Encourage students to cut out the reproductions (carefully) so that they can actually lay all the pictures down beginning at one corner of the graph.

Making the Lesson Harder

Provide less direction for creating the straight-line graph; instead, suggest the students use graph paper to create a graphic display of the good reproductions. Let small groups work together to see what they can come up with on their own. Give everyone credit for their creations, even if they are not what you may have been expecting. Assign *Extension: Paint By Numbers,* which begins with the dimensions in numbers.

Examining the pictures in light of the measurements led to a rich mathematical discussion about fractions in this class.

When asked whether the 5″ x 7½″ (Picture D) was a good reproduction, Carla quickly said, "It's going to be on the distorted side." "Why?" "You can't divide those two equally by some number."

I then asked them to try to show me that 5:7½ was not proportional. As usual, Marta resorted to dividing but couldn't successfully divide 5 into 7½. Carla suggested that she think in terms of money. I encouraged them to use chips so that they could "see" the numbers as they thought through the process.

Marta wrote "$1.00" on seven of her chips and then "$0.25" on two of them (rather than "$0.50" on one—clearly a direct link to her thinking about money). She doubled both amounts to get $10 on one side and $14 plus four $0.25's on the other side.

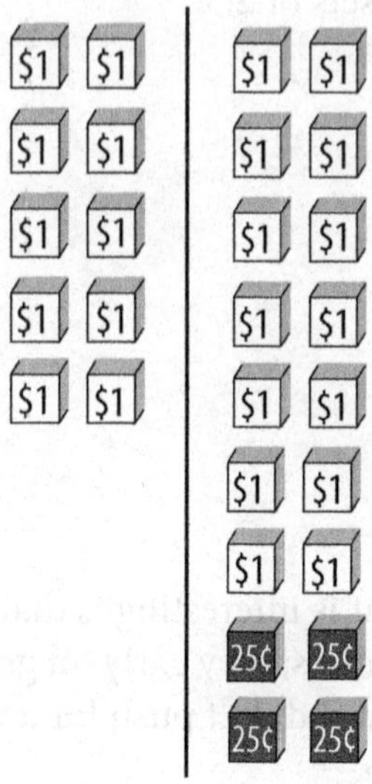

She changed her four quarters to one whole and reorganized her chips, trying to keep groups of 2's and 3's.

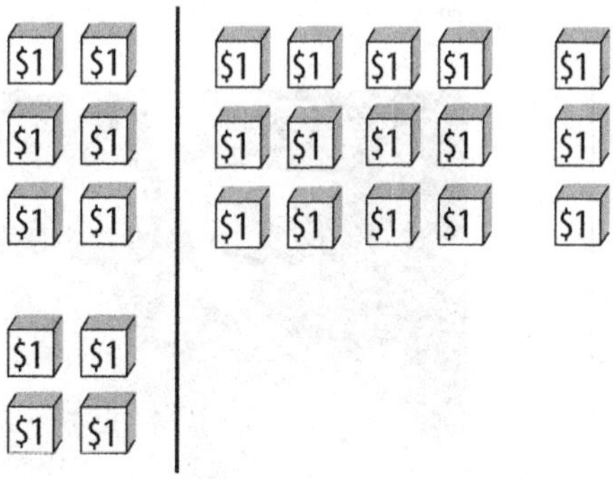

continued on next page

continued from previous page

I asked what she had. On one side she had grouped 10 into two groups of 3 and two groups of 2. For the other side, she commented, "I have four sets of 3 and three sets of 1."

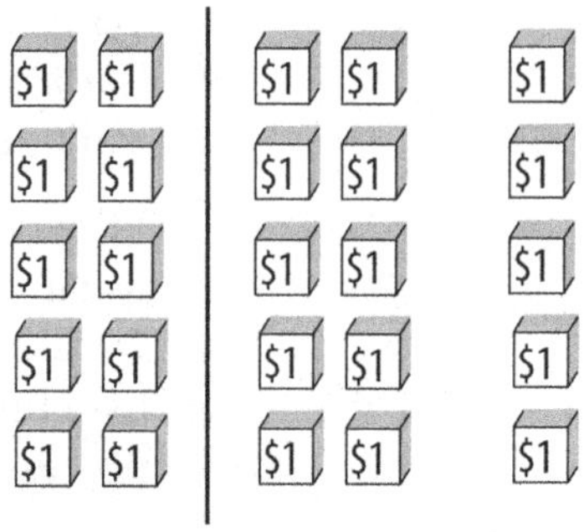

I asked whether there was another way to group the sets of numbers. She pondered and then reorganized them again. "What do you have now?" She had reordered the left side into five groups of 2. "I have five sets of 2 and five sets of 1 on this (right) side."

I finally asked her to get up and look at her chips from a different angle (because from my angle, it was obviously five groups of 3). Suddenly she saw what she had. Once she made her five piles, she lined up each set of 3 with each set of 2 so that she could clearly see that she had five sets of 2:3.

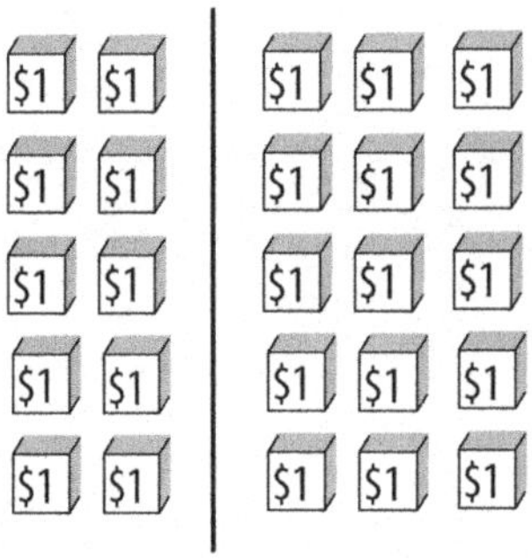

What is interesting is that both Carla and Marta worked with whole numbers; they early on got rid of the 7½ and never went back to it. At this point, I didn't push for a number by which we might multiply 2:3 to get to 5:7½.

Before ending for the day, the women charted the "good" pictures on their large chart. After they connected the stars, I asked what they saw (thinking they would say "a straight line." Marta quickly pointed out, "You have a chart!"

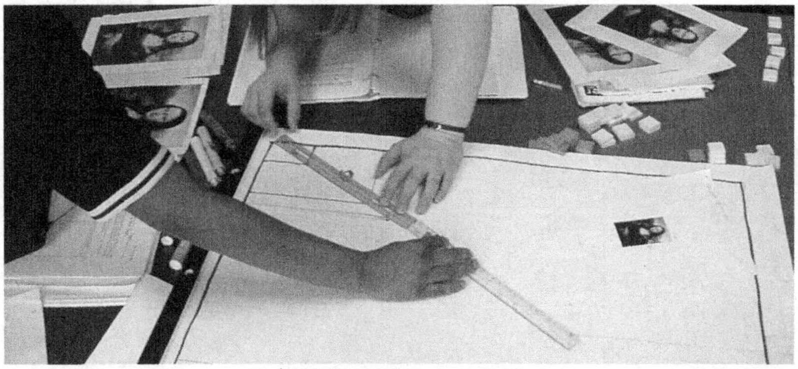

With a group of teachers, the authors discovered yet another way of checking to see whether each picture was in the same ratio as the original.

One person realized that the height of her picture was 1.5 times more than the width. To check, she folded the width so the bottom of the picture was flush with the side of the picture.

She then folded the top of the picture down, where it landed midpoint on the newly created triangle.

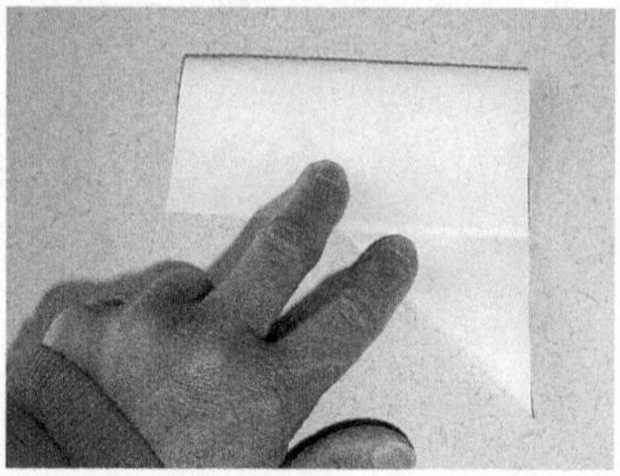

She creased the paper where the top edge landed. Finally she folded the triangle in half, directly on the crease, illustrating that the top third of the side was one-half of the width.

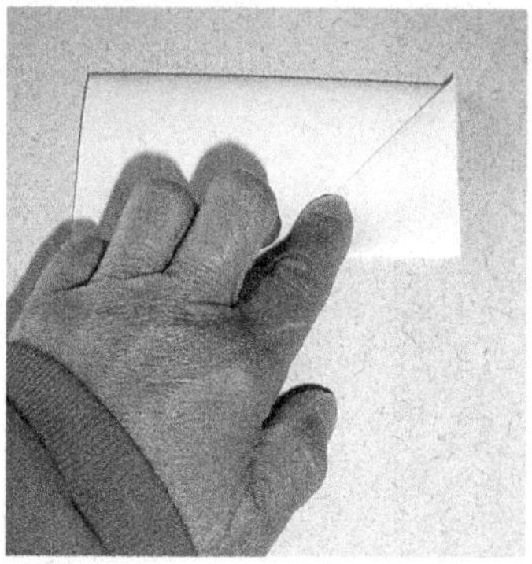

When examing the folds, she demonstrated that there were two "chunks" across the width, and three similar "chunks" along the length. We all saw the ratio of 2:3.

Donna Curry
Bailey Evening School, Bath, Maine

FACILITATING LESSON 6

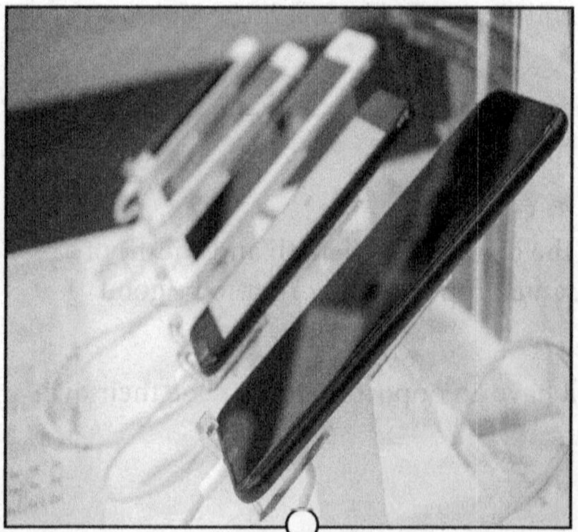

Redesigning Your Cell Phone

How do proportions affect design choices?

Synopsis

In *Lesson 5*, the dimensions (length and width) of the good *Mona Lisa* reproductions were kept to a constant ratio (2:3). In this lesson, students will have a chance to apply that idea by redesigning cell phones. Students examine cell phones (their own or those of classmates) and then design one with dimensions twice as large and another one with dimensions that are halved.

1. Students work individually to draw a cell phone actual size, and then larger and smaller versions.
2. Then they compare what happens to the area of the two newly sized cell phones.

Objectives

- Measure and draw rectangular shapes that are **similar** to one another
- Practice using decimals to describe fractions of a centimeter (to the nearest 0.1 cm)
- Contrast changes to area vs. perimeter when length and width are doubled or halved

Materials/Prep

- 0.5 cm grid paper or copies of *Blackline Master 16*, four pages per student
- Calculators
- Rulers marked in centimeters and millimeters
- Scissors
- Tape

Heads Up!

This activity could be done with any size cell phone. The lesson assumes students have experience with finding areas, covered in the *EMPower* unit *Over, Around, and Within: Geometry and Measurement* and with decimals (tenths), covered in *Split It Up: More Fractions, Decimals, and Percents.*

Opening Discussion

Begin by asking students to recall how the dimensions (length and width) of the *Mona Lisa* copies were kept in the same proportion in order to be good reproductions.

Explain that in this lesson, students will have an opportunity to create their own proportional objects.

Activity 1: Three Cell Phones

Distribute 0.5 cm grid paper or *Blackline Master 16,* four sheets per person. If students are using *Blackline Master 16,* have tape and scissors on hand so students can construct a larger piece of grid paper for the large version of the phone.

Ask students to read the situation about redesigning a cell phone (*Student Book,* p. 72). Allow time for everyone to draw their phones to size, labeling the length and width of the phone itself, as well as that of the display window. Have pairs check and verify each other's measurements to the nearest tenth of a centimeter. (Ascertain first that everyone knows how to read the ruler accurately to the nearest tenth of a centimeter.)

Then students draw the large-screen and compact-size versions of the phone according to the specifications, labeling the outside dimensions and the display window for each version.

Observe whether students are able to label in decimal form.

If a photocopy machine is available, ask students who finish their work quickly to make the following photocopies of their actual calculators:

- A copy that is true to size
- A copy that doubles the phone's dimensions
- A copy that halves the phone's dimensions

These students will then share with the class which buttons they pushed on the copying machine to double or halve the size of the actual phone's dimensions. This exercise connects percents to halving and doubling because copy machine ratios are usually given in percents.

Ask students to record their measurements and make observations about which ratios are equal on p. 73 of the *Student Book.*

Activity 2: Material for a Cell Phone Screen Protector

Direct students to read the situation presented in *Activity 2* (*Student Book*, p. 74). Encourage students to estimate the front face of each of the phones first before doing any actual measuring or calculations. Make sure everyone is clear about the difference between a centimeter and a square centimeter.

Ask:

Did anything surprise you as you worked through this activity?

Can you explain why the surface area is not doubled or halved when the length and width are doubled or halved?

Summary Discussion

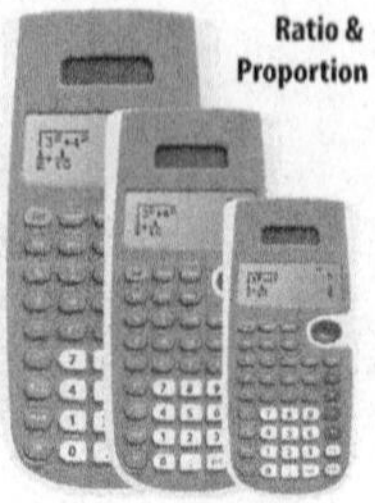

Review some of the tools that students have used to determine proportionality, including drawing pictures to show a pattern of ratios, multiplying or dividing by the same amount, and drawing a line graph to confirm that ratios are in balance.

Refer students to *Reflections* (*Student Book*, pp. 141–144). Have them explain in their own words what happens to the area of an object when you double the dimensions.

Practice

Similar Rectangles, p. 76
Helps students make connections between ratios and similar shapes.

Graphing Cell Phone Dimensions, p. 79
Extends the lesson by asking students to graph the phones they drew in *Activity 1*.

Shaping Up! p. 80
Students investigate the relationship between the areas of similar triangles, one of which has sides that are double the length of the other's.

Squaring Off, p. 81
Students investigate the relationship between the ratios of the length vs. the area of a square when the sides are doubled.

Extension

Extension: Perimeter, Surface Area, and Volume, p. 82
Extends the lesson to include a look at what happens to volume when three-dimensional shapes are enlarged.

Test Practice

Test Practice, p. 84

Looking Closely

Observe whether students are able to

Measure and draw rectangular shapes that are similar to one another

Students may draw by tracing the phone once and then again rather than first measuring the length of the original and then doubling the amount. If students draw one length and then add another equal length, ask them to check their accuracy by measuring the total length of the doubled size and comparing that to twice the length of the original.

Observe but do not intrude as students try to figure out where to place the phone screen in the new sizes. They may forget to consider that the distance between the edge of the phone and the screen must be doubled as well. If they draw the screen too close to the edge, they should notice either that the screen is off-kilter or that it is much larger than expected. Ask them to determine for themselves whether the screen is proportional to the rest of the phone.

Practice using decimals to describe fractions of a centimeter (to the nearest 0.1 cm)

The size of the original cell phones your students use will vary. Encourage your students to use the actual dimensions of the phone they use rather than rounding up or down to whole centimeters. This practice with decimals will prepare them for the next set of lessons that will involve working with decimals in the hundredths and thousandths places.

Contrast changes to surface area vs. perimeter when length and width are doubled or halved

Students may predict incorrectly the size of the area when the object is doubled or halved, but they should be able to articulate the pattern once they take the actual measurements and look for a pattern. They should notice that when doubling the dimensions of the phone, the perimeter doubles but the area quadruples; when halving the dimensions of the phone, the perimeter is half of the original size, but the area is only one-fourth of the original.

Encourage students to draw a picture to illustrate why this is so. If students do not understand why the area is four times as large when the length and width are doubled, you may want to show them one way to illustrate this:

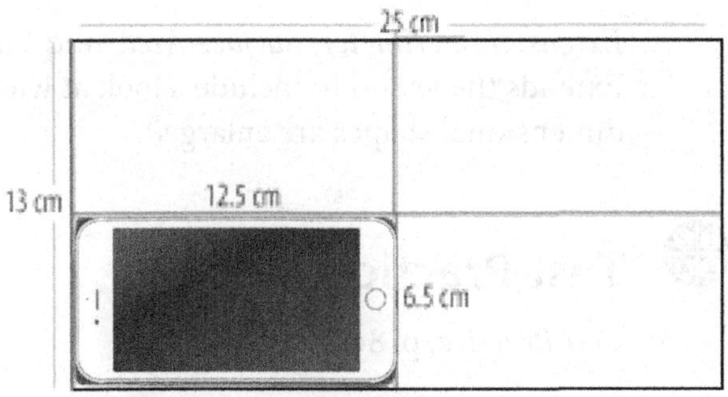

WHAT TO LOOK FOR IN *LESSON 6*	WHO STANDS OUT? (LIST STUDENTS' INITIALS)			NOTES FOR NEXT STEPS
	STRONG	ADEQUATE	NEEDS WORK	
Concept Development • Reasons proportionately to determine the correct size for a cell phone with dimensions twice the size of the original phone and one with dimensions half of the original's size • Can determine the surface area and perimeter of different sizes of rectangular solids				
Expressive Capacity • Accurately describes the relative relationship of the width to the length of the three phone sizes • Can explain the pattern for surface area and perimeter as sizes are doubled or halved				
Use of Tools • Accurately measures to the nearest 0.1 cm with a ruler • Uses number relationships to verify proportionality • Can draw a shape to scale using a ruler and grid paper				
Background Knowledge • Draws on past experience with perimeter and surface area				

Rationale

This lesson is designed to give further practice in applying the idea of equal ratios in visual contexts. Students extend their thinking to the notion that doubling the length and width does not necessarily mean that every aspect of an object is doubled.

Math Background

In this lesson, surface area and perimeter are reviewed. If students seem totally unfamiliar with these concepts, you may want to look at the *EMPower* unit *Geometry and Measurement: Over, Around, and Within.*

The notion that doubling the linear dimensions of a shape quadruples the area is an important mathematical concept that helps students make connections to other geometric ideas, such as 3 ft. = 1 yd., but 9 sq. ft. = 1 sq. yd. While the perimeter and dimensions scale up and down proportionately, the areas do not. Area does not grow in proportion to length or width.

A **table** displaying the perimeters and areas of mathematically similar rectangles:

Width (cm)	Length (cm)	Perimeter (cm)	Area (sq. cm)
3	5	16	15
6	10	32	60
12	20	64	240
24	40	128	960

A set of **equations** showing the relationships:

Perimeter

Original $P = 2(l + w)$

If the dimensions are doubled, New $P = 2\,(2l + 2w) = 4(l + w)$.

$$\frac{\text{New } P}{\text{Original } P} = \frac{4(l + w)}{2(l + w)} = \frac{4}{1}$$

Area

Original $A = lw$

If the dimensions are doubled, New $A = (2l)(2w) = 4lw$

$$\frac{\text{New } A}{\text{Original } A} = \frac{4lw}{lw} = \frac{4}{1}$$

The rectangle is the shape of interest here, but the relationship holds for other two-dimensional shapes as well. Doubling the sides of a triangle, hexagon, octagon, or any other polygon results in a quadrupled area.

Context

Direct proportion applies to many aspects of design when looking at relationships between linear features. However, it does not apply to all aspects (such as the manner in which area and volume increase). Some other situations to explore follow:

- How does doubling the linear dimensions of a rectangular plot of land, such as a city block, affect the number of houses that can be built? How does it affect the length of the sidewalk around the block?
- How much carpet will it take to cover an 8′ x 10′ room as opposed to a room twice as long and twice as wide?

Facilitation

Students may at first consider this a simple lesson because for most, the idea of drawing a rectangle that is twice as large or half the size of another is not too challenging. However, the activity quickly becomes more difficult when they have to place the screen and an app icon proportionately in their new rectangle. Moreover, they may find it quite challenging to figure out the amount of surface area needed for the new sizes of cell phones. Encourage students to draw pictures to visualize any patterns they have difficulty understanding.

Making the Lesson Easier

If students are struggling with decimal amounts, you may choose to use fractions (6½, 12½); or you may want to round the width to 6 cm and the length to 12 cm, numbers that are easier to halve. If you do, you may also need to omit working with the screen and the app icon of the half-size cell phone because the numbers may become too unwieldy for students.

Making the Lesson Harder

Students may want to look for patterns in the surface area vs. perimeter. Encourage them to think about the perimeter vs. surface area for a cell phone that is three times as large as the actual one, then four times as large, then five times as large. If they create a table of numbers, they can begin to see a pattern for both the perimeter and the surface area. From this pattern, they can predict for other sizes, both smaller and larger than the original.

Students who need a challenge may want to consider what happens to shapes when volume is taken into consideration. Suggest they imagine that not only are the length and width of the phone doubled (or halved), but also the thickness. Ask them first to predict what they think will happen, then to test their predictions by doing actual drawings and measuring the results. They should also be able to create a table of equal ratios of volume sizes vs. perimeter vs. area and make general statements about what happens to each when the linear dimensions change. Assign *Extension: Perimeter, Area, and Volume*, p. 82.

Encourage students to explore similarity with shapes other than rectangles. Assign *Practice: Shaping Up!* (*Student Book*, p. 80).

LESSON 6 IN ACTION

When using rulers, measuring issues often come up in class. The original lesson from the first edition of EMPower had students measuring and resizing calculators (not cell phones). This is why the lesson in action below refers to the Casio calculator. One teacher wrote about her experience.

Because we were limited in the types of rulers we had, students measured the Casio calculator in inches. They decided that the width was 2¾ inches (including the protective cover) and the length 5 inches. Two students first drew a length of 5 inches, then added another 5 inches. Next they drew 2¾ inches. I asked how wide the new calculator was. At first they were stumped. I suggested that they think of the measurement in monetary terms, and the students were able to figure out that the calculator was 5½ inches long. I then suggested they use the ruler to check whether their new calculator size made sense.

One student measured and determined that the width of the display window of the original calculator was 2 inches. Because she was using grid paper, she was able to center her window. She measured the distance between the side of the calculator and the side of the window. She figured it to be ⅜ inches wide on either side. I asked her whether that made sense (I could readily see that 2 + ⅜ + ⅜ = 2¾, but I wasn't sure she could.). She said that she didn't know because she did not really know how to read the little lines on a ruler.

Because students stated that they weren't comfortable with "those little lines," I decided to end the lesson by working on the ruler. We used a larger scale and first divided an inch in two, then each of those halves in half again. Eventually we got to 16ths. One student commented that it was interesting to note that every time we cut a fraction in half, the bottom number got twice as big. Yes!

Donna Curry
Gardiner Adult Education Program, Gardiner, Maine

Facilitating Interim Assessment: Checking In

Can people apply what they have learned?

Synopsis

In *Lessons 1–6*, students learned some basic tools to use in reasoning with ratios. To scale up or down, they drew pictures, looked for patterns, used the property of equal fractions, and created tables and graphs. This session is designed to see how well they can use these tools in a variety of situations. Students first work together on a project and then work individually on a series of three tasks.

1. Students work in triads to redesign a puzzle.
2. Individually, students solve two other open-ended tasks and a *Test Practice.*
3. A discussion on *Unit Goals* and personal goals reinforces the scope of the unit.

Objectives

- Create sets of equal ratios
- Use ratios to make comparisons, predictions, and decisions
- Write and interpret ratios in various formats

Materials/Prep

- Calculators
- Rulers marked in inches
- Scissors
- Tape
- Recipes (preferably brought in by students) or copies of *Pancake Recipe* (*Blackline Master 12*).
- ¼-inch grid paper (*Blackline Master 18*), three copies per person

Prepare one *Interim Assessment: Checking In*, pp. 153–158, per student.

Prepare one set of *Supermarket Flyers 1–3* (*Blackline Masters 1*, *2*, and *3*), per student.

Prepare two copies of *A Puzzling Situation* (*Blackline Masters 13*) for every team of three students.

Opening Discussion

The *Interim Assessment* is a chance for students to synthesize what they have learned in *Lessons 1–6*. Review by asking:

What are some of the tools for thinking about ratios you have been using over the past six lessons?

What are some things to look for to determine whether a set of ratios is equal or not?

Explain to the class that they are going to have an opportunity to use the different tools to solve several problems today.

The Interim Assessment: Checking In

Distribute one copy of the *Interim Assessment* to each student. Explain that there are four tasks. They are to do the first task in teams and the last three tasks individually.

Task 1: A Puzzling Situation

Assign teams of three students and give each team two copies of *A Puzzling Situation* (*Blackline Master 13*). Refer students to *Task 1: A Puzzling Situation,* p. 153, and distribute rulers, scissors, tape, and ¼″ grid paper (*Blackline Master 18*).

One team member neatly cuts the six pieces out of one puzzle, leaving one puzzle page as reference. Then each team member. chooses two pieces and creates and cuts out new versions. The new pieces, when put back together, will form a square.

Once students are clear on the directions, tell them that their goal is to create a new, larger square. When they have done so, ask them to answer the questions pp. 153–154.

When everyone has completed the assignment, ask students to share what they discovered:

What was easy about the activity?

What was challenging?

Task 2: Recipe Project

For this task, have each student use a recipe that they brought to class (or one that they can recall from memory because they will be working with the ingredients only). If students do not have their own recipes, you might have them use the *Pancake Recipe* (*Blackline Master 12*).

Ask students to use the chart in their *Interim Assessment*, p. 154, to write out the ingredients for their recipe in column 1. Then they are to create variations of the recipe in each of the other columns.

Finally, they will need to choose two of the ingredients and show that the different amounts are all in proportion. They will use those two ingredients to create a graph of the ratios, again to show that the two ingredients remain in proportion to one another.

Task 3: How Low Can You Go?

Students will need one copy of each of the three supermarket flyers (*Blackline Masters 1–3*) for this task. They are to look at the costs of each of of the items and decide where they would shop to get the lowest price. Their goal is to get the lowest price on *each* item, not the overall lowest cost at one store. Students may work with a partner or on their own.

Once they have determined the lowest price for each item, students explain where they would buy the item and how much they would pay.

Task 4: Test Practice

Students complete these problems individually.

Summary Discussion

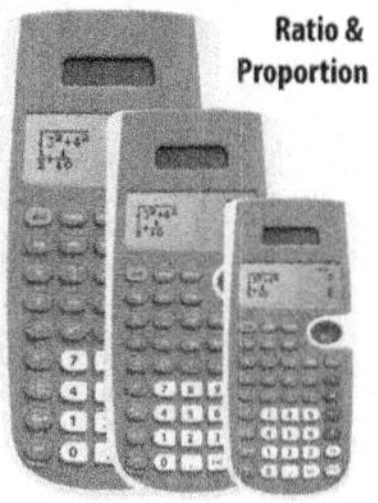

Ask students how they felt about the *Interim Assessment*:

What did you feel particularly good about?

Were there any particular issues that you would like addressed before moving forward?

To close, refer students back to the *Unit Goals* on p.5 for a discussion about where they are now in relation to those goals.

Looking Closely

Observe whether students are able to

Create sets of equal ratios

By now, students should be able to create sets of equal ratios using friendly numbers. They may refer to pictures or make a table and use the property of equal fractions. They should be able to also determine when two sets of ratios are *not* equal.

Use ratios to make comparisons, predictions, and decisions

Can students see the purpose of creating equal sets of ratios? Do they see that creating a ratio for a smaller set allows them to predict what the larger set might look like?

Are they able to compare two sets of ratios to determine whether they are equal? They should be able to look at two pairs of ratios and determine which of the two represents a lesser amount (for example, a cheaper price).

Write and interpret ratios in various formats

Students should be comfortable writing ratios in fraction form (2/5) or with a colon (2:5), as well as in words (2 for $3, 5 out of 6). Students should be able to recognize these different formats and be able to change formats in order to make comparisons. They should also realize by now that a unit price is also a ratio. For example, they should understand that $1.99 for a box of crackers is the same as 1 for $1.99 or 1/$1.99.

WHAT TO LOOK FOR IN *INTERIM ASSESSMENT*	WHO STANDS OUT? (LIST STUDENTS' INITIALS)			NOTES FOR NEXT STEPS
	STRONG	ADEQUATE	NEEDS WORK	
Concept Development • Understands how to create equal proportions given a ratio • Uses ratios to make predictions • Can make comparisons among a set of ratios to determine which is cheaper				
Expressive Capacity • Can explain clearly why two sets of ratios are in proportion • Describes the reasoning behind decisions made based on ratios				
Use of Tools • Uses pictures or objects to show why two ratios are equal • Uses property of equal fractions in creating sets of equal ratios • Uses graphs to confirm that sets of ratios are in proportion				
Background Knowledge • Uses prior experience with shopping to determine what strategy to use to make comparisons				

Rationale

This lesson is designed to show students what they have learned about ratio and proportion. The assessment includes a variety of situations in which students need to use the tools they have learned thus far.

Math Background

This lesson concludes proportional work with friendly numbers. The next set of lessons will expose the student to less friendly numbers and therefore new tools to handle those numbers.

Facilitation

This lesson is designed to be an assessment of what students can do, so try to stay in the background and let the students do all the work. If students cannot complete an activity or task, make notes on the types of questions they ask. Probe to find out what they are thinking when they have difficulty with a problem.

Making the Lesson Easier

If this is a new experience for any of your students (who may have come in after the *Initial Assessment* was given), you may want to keep the problems displayed so that students will not feel overwhelmed. Read each task aloud for students just to be sure there are no issues with readability. Then ask them to decide whether they can or cannot do each of the tasks shown.

Two students worked on the ratio of rice to water, based on a recipe that called for one cup of rice for every two cups of water. Notice how each approached the graph.

To graph the ratios, the first student, Callie, used a format similar to the *Mona Lisa* graph from *Lesson 5*. In her explanation, she said that she made each square equal to one cup. She said that the water was shown on the bottom of the chart and the rice on the top. To make her chart, she "went up 1 and over 2 for 2 cups of rice to one cup of water."

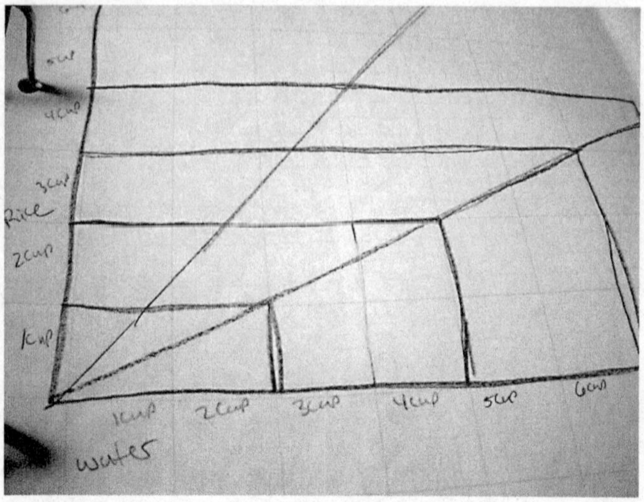

Mayla, on the other hand, created a bar graph. She used the grid paper appropriately and labeled her graph.

The teacher commented:

"I asked Mayla what she was thinking as she created her graph. She said, 'I marked 4 people at the bottom to show a group of 2 (rice to water); all the way up to 4 rice and 8 water for 16 people.' She went on to add, 'The other chart [*Mona Lisa*] is not a bar chart, but mine is.'"

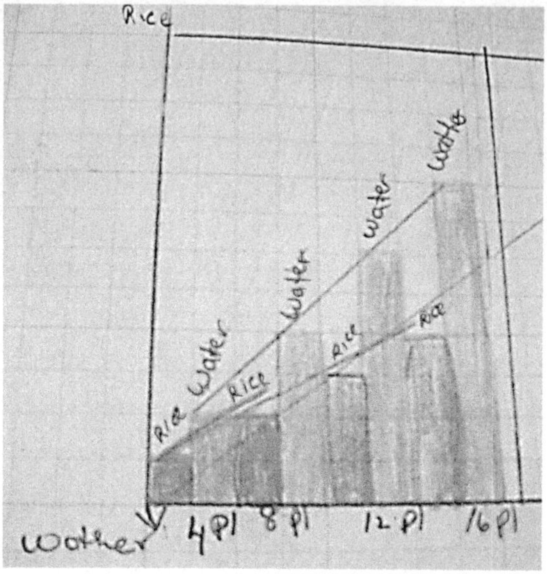

Donna Curry
Bailey Evening School, Bath, Maine

FACILITATING LESSON 7

Comparing Walks

What's our average walking speed?

Synopsis

This lesson focuses on **speed** and enlarges the language of ratio by introducing **unit rate**.

1. Students discuss the concept of rates and provide examples of rates used in their lives.
2. In small teams, students calculate their personal walking speeds.
3. Class data and strategies are compared. Students conclude who in the class has the fastest walking speed and who has the slowest.
4. The form a/b is now interpreted as the quotient ($a \div b$). Division is introduced as a way to establish the unit rate so the class can reach agreement on the fastest walker.

Heads Up!

From this point on, all lessons will use more unwieldy numbers, so calculators will be very useful. However, estimation and mental math skills are always used to verify the reasonableness of the answers.

Objectives

- Describe and quantify speed as the relationship of distance and time
- Develop a strategy of finding the unit rate for a/b, using division as another way to restate a rate or ratio

Materials/Prep

- Calculators
- Easel paper
- Masking tape
- Stopwatches
- Measuring tapes

Opening Discussion

Display the word "rate" and open with a question such as this:

What are some rates you come across in everyday life?

Add students' examples of rates in their lives. Students might mention some of the following:

- Pay rate
- Subscription rates
- Tax rate
- Mortgage rates

Ask:

A rate always relates two quantities. What are the two quantities in each of these rates?

Probe for labels to go along with examples.

For instance:

- Pay rate can be dollars per hour ($18/hr.)
- Subscription service ($49.99/month)
- Sales tax rate is cents per dollar ($0.06/$1.00)
- Real estate tax rate is a certain amount per $1,000 assessed house value
- Mortgage rate is interest per year on money borrowed (a 4.5% interest rate means $4.50 interest on every $100 borrowed)

Summarize:

Today's lesson is on speed, a type of rate. As you calculate speed, keep in mind the two quantities that determine speed.

Activity 1: Team Walking Speeds

The purpose of this activity is for students to determine and compare the walking rates of each person in a small group. If possible, go outside to do this activity or use a long hallway.

Ask for a volunteer to read the directions for *Activity 1* (*Student Book*, p. 86), and then form teams of three or four students. Remind students to choose someone to write up information and present the team's information to the class.

As the teams work, observe

- How people are estimating or measuring distances;
- What they are deciding to include and why;
- How they are including everyone in their work; and
- How they are keeping track of their information.

There are numerous productive strategies. Some students may decide to keep the distance constant (for example, marking off 20 feet) and focus on the variation in time. Others might clock everyone for a constant time (say, 15 seconds) and record the various distances. And others might calculate the unit rate.

Heads Up!

This should be an open-ended activity with students determining their own process for how to do the activity and their own units of measurements in a way that makes sense to them.

When all groups have presented their findings, ask:

Who from your team walked fastest? How do you know?

What is the average walking speed in your group? How do you know?

Activity 2: Class Walking Speeds

Now you want students to compare the rates of speed among all the teams.

Ask:

With the data we have, can you tell who walked the fastest in the class?

In most cases, students will find it easy to reason about who walks the fastest within their small team. A dilemma arises, however, when they compare the fastest people on each team if they have used different units of measure or different ways to compare.

Heads Up!

If this dilemma does not arise because everyone used the same method and units, present the following situation. Tell students that in another class,

- Team 1's winner, Kedian, walked 15 yards in 10 seconds;
- Jake, Team 2's winner, took 22 seconds to walk 100 feet; and
- The winner for Team 3, Jenna, walked around the room 5.5 times in one minute. The room is 80 feet around.

If students seem stumped, ask the class to decide on a common unit of measure for distance and time. Students convert the fastest walking rates of all the teams to the common unit of measure so they can make comparisons. Then they work in teams to fill in the table on p. 87 of the *Student Book*.

Explain that division is one way to interpret a rate or a ratio (a/b) and especially helpful when you want to compare unit rates or ratios. Ask everyone to use a calculator to divide the distance by the time. Be sure that students have the order correct, and take time to discuss what that quotient actually represents—the distance per one unit of time.

This method is used when you convert fractions to decimals. Think of any ratio in fraction form: 45 ft./10 sec. as 45 divided by 10, or 4.5 ft. per sec.; 100 ft. in 22 sec. as 100 divided by 22, or 4.5454 ft. per sec.

Heads Up!

Assign *Practice: Converting to the Unit Rate* if students need practice.

Summary Dicussion

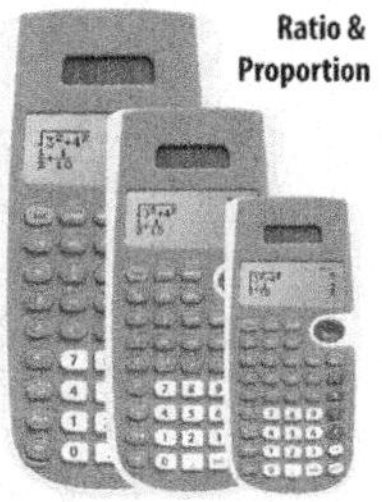

Close by asking:

What are two elements that make up speed?

What was the same about how the groups figured out the problem? What was different about how the groups worked on the problem?

Now remind students of the work they did at the stations from *Lesson 2: It's a Lot of Work!* You might want to refer specifically to some of the exact numbers used by students in their sampling. Ask:

How was this the same or different from establishing a sample rate of work in *Lesson 2*? For example, if someone peeled *X* potatoes in *Y* minutes, what would the unit rate be?

Refer students to *Reflections* (*Student Book*, pp.141–144) and have them explain in their own words how to find a unit rate and why it is a useful tool.

Practice

Rounding Decimals From the Calculator Screen, p. 88
For practice rounding decimal answers that appear on the calculator screen.

Converting to the Unit Rate, p. 89
For practice finding unit rate by dividing the numerator by the denominator.

Who Is the Fastest? p. 90
For further practice finding unit rates to make comparisons.

MPH, MPG, and Other Rates, p. 91
For practice rounding and estimating.

Extension

On the Treadmill, p. 92
For practice with speed, distance, and time.

Graphing Treadmill Distances versus Times, p. 94
For practice graphing rates.

Test Practice

Test Practice, p. 95

Looking Closely

Observe whether students are able to

Describe and quantify speed as the relationship of distance and time

This lesson is designed to have students experience the mathematics of speed by actually getting up and moving around. This kinesthetic activity leads to an important new tool for determining proportionality.

Are students able to record their times and distances? Do they know how to proceed with the task without guidance? Let students work together without your assistance to figure out which process they want to use. They do not need to consider which formula to use, but rather how to use ratios and sample size to develop a prediction about the walking speed of individual students.

Develop a strategy of finding the unit rate for *a/b*, using division as another way to restate a rate or ratio

Because students are now going to be looking at ratios with unwieldy numbers, it is important to give them a new tool to use. Although cross product is often used with more difficult numbers, division can be very useful in making direct comparisons between ratios.

Notice whether students are consistent in writing *a/b* vs. *b/a* so that their quotients are of the same ratios. Though it appears that there is only one number represented when writing a quotient for a ratio, there are actually two. Fifty yards in 40 seconds (a ratio of yards to seconds, 50:40) can be written as a quotient (50 divided by 40, or 1.25 yards per second), a ratio of 1.25:1. Be sure that students use labels to describe their quotients. This will help them determine whether the quotient they compute is reasonable.

WHAT TO LOOK FOR IN *LESSON 7*	WHO STANDS OUT? (LIST STUDENTS' INITIALS)			NOTES FOR NEXT STEPS
	STRONG	ADEQUATE	NEEDS WORK	
Concept Development • Relates speed to the idea of rates and ratios • Predicts for longer distances based on sample ratios • Makes comparisons of speed by creating similar units of measurement to compare				
Expressive Capacity • Describes the walking speed of different individuals • Writes a ratio as a quotient with labels • Can explain clearly how to determine the rate of speed				
Use of Tools • Uses pictures or objects to show that two ratios are or are not equal • Uses a mathematical rule (such as "multiply both amounts by the same number") to create a new ratio equal to a given ratio				
Background Knowledge • Applies knowledge from personal experience with rate, including speed				

Rationale

Until now, students have been working with fairly friendly numbers, so tools—such as pictures—and rules—such as multiplying by the same number—have been sufficient. In *Lesson 7,* students will no doubt face more difficult numbers as they collect the data. When the numbers get unwieldy, a new tool is needed. Although students should be encouraged to continue to use tools they learned in earlier lessons, they will need practice with the new tool.

Math Background

One of the biggest challenges in using quotients to describe ratio is deciding what that one number represents. Labels are critical to understanding exactly what the quotient means.

With unit rates using quotients, order *does* make a difference. For example, think about the *Mona Lisa* ratio from *Lesson 5.* If the ratio of the width to the height of *Mona Lisa* is written as 2 to 3, then changed to a quotient (0.667), the width is 0.667 times as large as the height, or 0.667″ wide to every 1″ of height. If the height (3″) is compared to the width (2″), then the quotient becomes 1.5″—meaning that the height is 1.5 times larger than the width, or 1.5″ of height per 1″ of width.

Facilitation

Students should have time to explore the task among themselves in their teams. Make sure that everyone is involved (keeping the teams small—three or four students each—will help) and can describe how individual rates of speed were determined.

Because it is important that students determine their own method of describing rates of speed, you may discover that they have different descriptions, depending on how each group tackled the project. This is an opportunity to share that people often describe rates in different terms; what is important is that comparisons can still be made among rates once rates are converted to similar terms. Again, let students work together to determine a common way to describe rate of walking speed. Ask guiding questions, if necessary, and give direction only if students seem to be confused.

Making the Lesson Easier

As you observe groups, pay attention to who may be struggling with difficult numbers. You may want to encourage these students to round to friendlier whole numbers to make comparisons within the teams easier.

Making the Lesson Harder

For those students who want a challenge, you might suggest that they develop yet another method for describing the rate of speed for class members. For example, if the class decided to describe the rate of speed in feet per second, these students might want to try to determine the rate of speed in miles per hour—a challenging number workout. Have calculators available.

Students in Holyoke, Massachusetts used the floor of the classroom and the hallways as places to conduct their investigations.

Some groups of students decided to keep the distance constant by marking off a given distance and recording the time it took for each member of the team to walk that distance. One group presented this data:

Team Member	Distance	Time
Allen	34 feet	8.5 seconds
Andy	34 feet	9.58 seconds
Jessie	34 feet	9.80 seconds

They wrote, "We all timed ourselves based on the same amount of feet, and our conclusion was that Allen was the fastest and Jessie was the slowest."

Another team settled on a different length as their constant distance:

D (ft.)	T (sec.)	D (ft.)	T (sec.)	D (ft.)	T (sec.)
32	20 Sue	16	10 Sue	8	5 Sue
32	20 Justin	16	10 Justin	8	5 Justin
32	32 Mikki	16	16 Mikki	8	8 Mikki
32	24 Everet	16	12 Everet	8	6 Everet

They wrote, "Sue and Justin are tied (for fastest), Everet is second, and Mikki last (slowest). Our team decided to walk 8 feet and write down how many seconds (time) it took us to walk 8 feet (distance)."

Finally, one team used estimating, rounding, and calculators to determine who walked the fastest.

continued on next page

continued from previous page

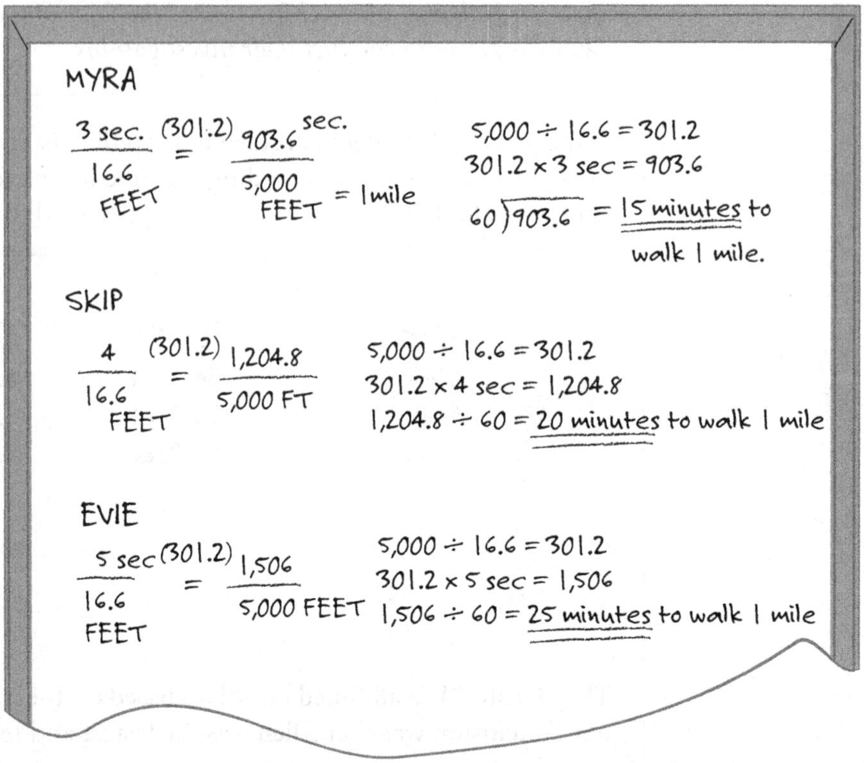

Although each group of students was readily able to compare rates within their own team, the challenge came when the entire class looked across the teams. At this point, students clearly saw the usefulness of using one common unit rate to make comparisons.

Tricia Donovan and Glenn Yarnell
HALO Center, Holyoke MA

In another class, the teacher described what happened as she observed one team working together.

I observed the group as they first determined that they could do the project in the room rather than in the halls, so they decided to time each other using the length of the room. As I listened, I heard them comment first that the distance to be walked was 24 feet, 8 inches. I anticipated this would get them into cumbersome calculations. But after some negotiating among themselves, they decided 25 feet sounded a lot better! Once they timed each other, they sat together to figure out what to do next. One student was clearly in charge and was writing down all the facts.

continued on next page

continued from previous page

She first wrote, "Length we walked: 25 ft. Dan—10 sec.; Lynda—10 sec.; Tai—8 sec."

After some further discussion among the group, she added: "5,280 ft. = 1 mile; 60 sec. = 1 min.; 60 min. = 1 hour."

She then set up a ratio:

$$\frac{8 \text{ sec.}}{1{,}760 \text{ sec.}} = \frac{24 \text{ ft.}}{5{,}280 \text{ ft.}}$$

The students determined the missing quantity by looking across and seeing that 8 x 3 = 24. So, if they wanted to compare that to a mile (5,280 feet), then 3 x what number would give them 5,280? They divided 1,760 by 60 seconds to figure that Tai, the fastest, would walk a mile in just a little more than 29 minutes. The student writing on the board wrote, "1/2 hour to walk 1 mile."

We talked about whether that made sense. I then asked them whether it would have worked out the same if they had written the ratio differently, as follows:

$$\frac{8 \text{ sec.}}{1{,}760 \text{ sec.}} = \frac{?}{5{,}280 \text{ ft.}}$$

They were not quite sure, so we worked it out together. The students reasoned that they had to multiply 220 across the top and bottom. After figuring it out, we talked a bit about why the ratios were the same.

Donna Curry
Gardiner Adult Education, Gardiner, ME

FACILITATING LESSON 8

Playing with the Numbers

What happens when numbers are not so user-friendly?

Synopsis

In this lesson, students focus on the numerical relationships within a proportion equation, regardless of context. Using the cross-product property and other tools, students examine and create a true proportion equation and a false one. When the numbers get more complicated, they use estimation prior to calculating more exact answers for the missing number in a proportion.

1. The whole class looks within and between ratios to examine relationships and to test out a final new tool, the cross-product property.
2. Student pairs create and check proportion equations (one true and one false).
3. Given any three numbers, student pairs find the fourth number to complete a proportion, first estimating a reasonable amount and then calculating the answer.
4. The class summarizes what they have learned and what remains confusing.

Objectives

- Use the cross-product property of proportions as another tool to check that two ratios are equal
- Distinguish between a true and a false proportion equation
- Use estimation to predict for more complicated numbers in proportions
- Determine the missing number in a proportion

Materials/Prep

- Calculators

Copy *Card Set 1: Friendly Numbers* and *Card Set 2: More Challenging Numbers (Blackline Masters 14* and *15)* and cut them into two sets of four cards for each pair of students.

Opening Discussion

Begin by asking students to reflect on ways to know for certain whether two ratios are equal. Write on the board:

$$\frac{400}{25} = \frac{2,000}{125}$$

Ask:

How could you test whether this proportion is true?

Listen to ensure that students are looking both within each ratio (400 is 16 x 25 and 2,000 is 16 x 125) as well as between ratios (400 x 5 = 2,000 and 25 x 5 = 125). Or by using division, they may see that 400 ÷ 25 = 16 and 2,000 ÷ 125 is 16.

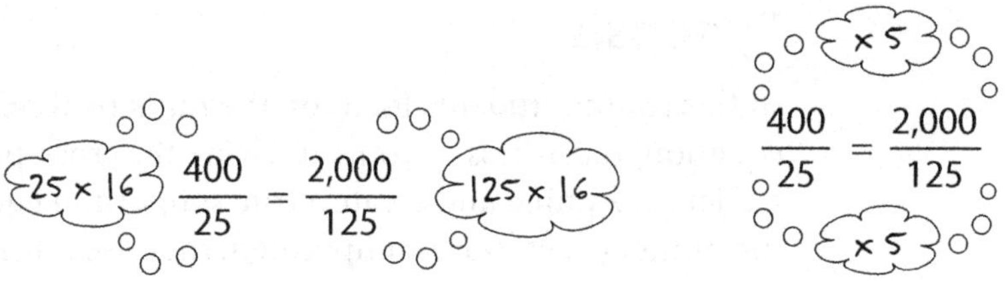

$$25\overline{)400}\ \ 16 \qquad 25\overline{)2,000}\ \ 16$$

If necessary, prompt with questions such as these:

What is the relationship between 400 and 2,000? Between 25 and 125?

What is the relationship between 400 and 25? Between 2,000 and 125?

What is 400 divided by 25? What is 2,000 divided by 25?

Introduce another interesting relationship: The product of 400 and 125 is 50,000, and the product of 2,000 and 25 is 50,000. In other words, the cross products are equal.

Activity 1: Creating and Checking Equal Ratios

The purpose of this activity is to introduce and practice using the cross-product property, along with other tools, to consider whether two ratios are equal.

Refer students to *Activity 1: Creating and Checking Equal Ratios (Student Book, p. 98)*. Working in pairs, students complete two tasks. First, they create a true proportion equation and prove that the ratios are equal in two ways, one of which is using the cross-product property. Second, student pairs agree on a proportion statement that is not true, using the cross-product property and another method to show that it is not.

When students are done creating a true and a false proportion, ask each pair to display the proportion equations they created, placing them in the correct category (true or false). Then ask the class to examine all the ratios, verifying whether each has been classified correctly. Allow time for discussion until consensus is reached about the correct classifications.

Students' contributions could look something like this:

True	False
$\frac{400}{25} = \frac{2{,}000}{125}$	$\frac{25}{125} = \frac{2{,}000}{400}$
$\frac{25}{125} = \frac{400}{2{,}000}$	$\frac{400}{25} = \frac{125}{2{,}000}$
$\frac{125}{25} = \frac{2{,}000}{400}$	$\frac{25}{2{,}000} = \frac{125}{400}$

Summarize:

You have several ways to reason whether two ratios are equal. You will, of course, tend to use the one that is easiest for you.

Activity 2: What Is the Fourth Number?

Distribute two sets of number cards (*Card Set 1: Friendly Numbers* and *Card Set 2: More Challenging Numbers* from *Blackline Masters 14* and *15*). Refer to *Activity 2: What Is the Fourth Number? (Student Book*, p. 99).

Card Set 1: Friendly Numbers

Tell student pairs to examine the number cards of *Card Set 1* (5, 10, 20, and a blank). Ask them to set up a proportion in this form:

$$\frac{\square}{\square} = \frac{\square}{\square}$$

They can choose to place the cards in any order they wish. If no one mentions it, point out that one of the cards is blank.

Once students have placed their cards on the proportion equation, tell them their task is to write on the blank card the fourth number that would make the proportion equation true. They should be prepared to verify that it is true, using the cross-product property and one other method.

Bring the class together. Ask each group to display all their final proportions for the numbers from *Card Set 1*.

Heads Up!

Although they will not all surface, there are eight possible proportions that could be formed and three possible solutions for the missing number, depending on the proportions equation (2.5, 10, or 40).

Allow time for class discussion to settle any disagreement on the missing number solutions.

Card Set 2: More Challenging Numbers

Until now, students have been working mostly with friendly numbers when considering proportions. This task is designed to push them to handle more difficult numbers. Make calculators available.

When they have come to agreement on *Card Set 1*, tell student pairs to examine the cards from *Card Set 2* (251, 76, 123, and blank) and to set up a proportion in this form:

$$\frac{\square}{\square} = \frac{\square}{\square}$$

Again they may choose to place the cards in any order they wish.

Once they have placed the cards on the proportion equation, tell them they are first to *estimate* the fourth number—a ballpark figure—and to write the estimate on the blank card. Remind them that when numbers are unwieldy, it is always useful to estimate first. Ask:

Approximately what amount would make the proportion equation true?

Then students should figure out a more exact missing number, checking against their estimates. Encourage them to work together, use what they know, experiment with the calculator, and think about how multiplication and division are related. Allow time for exploration.

Bring the class together. Ask each pair of students to share their final proportions for the numbers from *Card Set 2*, displaying all the equations they have created.

Ask:

Are all of these true proportion equations? How do you know?

What is one way that you used to figure out a more exact answer?

How did estimation help?

Heads Up!

Students should be able to use different methods, including the unit rate that they learned in *Lesson 7*. Again there are eight possible proportions and three possible answers, depending on the proportion. Answers to the nearest hundredth are 37.24 (any estimate between 30 and 40 is reasonable); 155.09 (an estimate of about 150 would be reasonable); or 406.22 (somewhere between 350 and 450 would be a reasonable estimate).

Summary Discussion

Summarize by asking:

You now have several ways to reason whether a proportion is equal, one of which is the cross-product property. Are you convinced the cross-product property will always work? Why or why not?

Refer students to *Reflections* (*Student Book*, pp. 141–144) and suggest that they explain in their own words how to use the cross-product property to check whether two ratios are equal. Ask them also to explain how to use estimation when working with challenging numbers.

Practice

Relating Multiplication and Division, p. 102
For practice seeing the relationship between multiplication and division.

What Is the Missing Number? p. 103
For practice determining the fourth number in a proportion.

Estimate and Solve, p. 104
Provides practice using the cross-product property with more challenging numbers.

Patterns for Predicting, p. 105
For practice determining estimates based on relationships across ratios.

Story Problems, p. 106
For practice with estimating and determining actual amounts.

Extension

Bringing Proportions to Life, p. 107
Gives students a chance to bring meaning to the numbers used in *Activity 2*.

Test Practice

Test Practice, p. 108

Looking Closely

Observe whether students are able to

Use the cross-product property as another tool to check that two ratios are equal

Some students may have had extensive experience with the cross-product property and may be able to use this tool without any prompting. Other students may have been exposed to the tool before but did not understand its purpose or see its value, especially if they were able to use other strategies effectively in their proportional reasoning. In any case, encourage students to use the cross-product property to check their results, but to use other tools also for arriving at equal ratios.

Understanding *why* the cross-product property works is another issue and depends on algebraic proficiency. As a teacher, you should understand why it works. Look under *Math Background* on p. 115 for an explanation.

Distinguish between a true and a false proportion equation

What strategies do students use to determine whether a proportion is true? Check whether they are able to rely on strategies they already knew, as well as the new ones recently introduced.

Use estimation to predict for more complicated numbers in proportions

Are students able to use estimation in their initial analyses of the proportions? Expect to have to make quick decisions on rounding and estimating to make sense of the relationships. Can students round to numbers that relate to one another well and make it easier to estimate? For example, if the ratio were 251 to 123, can they see that it would be easier to round 123 to 125, rather than 120 or even 100, because 125 is half of 250? Or, using the unit rate concept, they might estimate the division of 251 by 123 as 2 and ask what number would divide into 76 to give the same result of 2 (a number between 30 and 40).

Are students able to estimate as they look across the ratios? Can they estimate in order to compare within a ratio? If students have difficulty with the numbers suggested, begin by using only whole numbers up to 100. This activity encourages students to build strong number-sense skills.

In this lesson, students must do a lot of thinking about how to estimate. This estimation is not about using a rule for rounding to the nearest whole unit, but rather about reflecting on the relationship between and among numbers and looking for some way to describe that relationship. To help students reason in this way, you may need to model the process. Verbalize your reasoning as you figure out relationships between a pair of numbers.

Students may want to skip the prediction part of the activity and go straight to finding the "right" answer. Remind them that they have to know how to predict to know whether their "right" answer is reasonable or not. Making predictions strenghtens students' number sense, rather than just rote memorization of tools and strategies.

Determine the missing number in a proportion

What strategies do students use to find the exact or almost exact proportion? Are they able to get close to the answer by doubling or multiplying, but unable to get to the exact number? Do they use the relationship between multiplication and division? Because these numbers are so difficult and result in answers with many decimal places, issues about rounding decimals will no doubt arise in the discussion.

Given the numbers 251, 76, and 123 from the example in *Activity 2* (*Card Set 2*), students could use several strategies.

For example, for this set-up, 76/123 = ?/251, if they look between ratios (the property of equal fractions), they could start by thinking, "What number multiplied by 123 would equal 251?"

They could look within ratios and ask, "What number multiplied by 76 equals 123?"

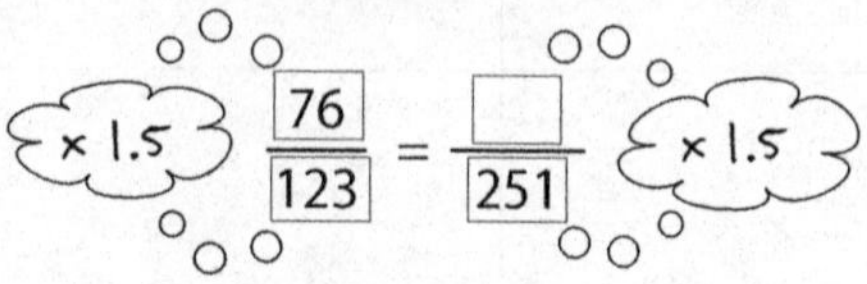

Using division (unit rate), students could start by thinking, "76 divided by 123 = 0.62. What number divided by 251 = 0.62?"

$$\frac{76}{123} = \frac{\square}{251} \qquad 251\overline{)\;?\;}^{\,0.62}$$

Using the cross-product property, they could ask, "76 multiplied by 251 = 19,076. What number multiplied by 123 equals 19,076?" Or "19,076 divided by 123 equals what number?"

$$\frac{76}{123} = \frac{\square}{251}$$

76 x 251 = 19,076

123 x ? = 19,706

WHAT TO LOOK FOR IN *LESSON 8*	WHO STANDS OUT? (LIST STUDENTS' INITIALS)			NOTES FOR NEXT STEPS
	STRONG	ADEQUATE	NEEDS WORK	
Concept Development • Organizes the four cards into a format that resembles a proportion • Rounds numbers based on a close relationship of the numbers • Predicts the missing amount based on a reasonable estimate				
Expressive Capacity • Explains why a proportion is true or false • Articulates why numbers were rounded the way they were • Explains why the prediction is a reasonable estimate				
Use of Tools • Uses at least two tools to show that two ratios are equal • Uses a quotient to check whether exact answer is valid • Uses cross product to check whether a proportion is true and to determine an exact answer				
Background Knowledge • Draws upon prior exposure to proportional reasoning with unwieldy numbers				

Rationale

This lesson is the only purely decontextualized one in the unit. Students focus on the numbers themselves. Up until now, students have been encouraged to "see" the relationship between and across ratios. The cross-product tool, although useful in many proportional situations, does not allow students to connect to the proportionality of the situation easily. Here, given extensive opportunities to play with proportion, they should be ready to try using the cross-product tool.

Math Background

Why there was no rush to the cross product as a proportional reasoning tool

The literature on fostering proportional reasoning includes numerous exhortations to mathematics educators to hold off for a while before introducing the cross-product property (see, for example, Cramer and Post, 1993; Langrall and Swafford, 2000; and NCTM 2000, p. 221). Two reasons are commonly given: (1) Cross-multiplication is sometimes overused (automatically applied to nonproportional situations) or misused (people scramble the numbers); and (2) Cross-multiplication distances people from the meaning in the situation. It is difficult to link the procedure with the mental operations of scaling up or down. For example, if 2 items cost \$5, it makes sense that 20 items cost \$50 because each new amount is 10 times as large as in the original situation. But it is hard to connect 2 multiplied by \$50 to 20 multiplied by \$5 in a concrete way.

Nonetheless, cross-multiplication is both efficient and universally applicable. This lesson encourages students to coordinate the use of cross product with the tools used in earlier lessons.

The mathematics of the cross-product property

Although in this unit the treatment of why the cross product works is more inductive—seeing it in a number pattern that occurs over and over—the more formal proof for the cross-product property is as follows:

$\frac{a}{b} = \frac{c}{d}$	Given.
$d \bullet \frac{a}{b} = d \bullet \frac{c}{d}$	Multiply both sides of the equation by the same number (*d*).
$\frac{ad}{b} = c$	Cancel common factor *d*.
$b \bullet \frac{ad}{b} = b \bullet c$	Multiply both sides of the equation by the same number (*b*).
$ad = bc$	Cancel common factor.

Conversely,

$ad = bc$	Given.
$\frac{1}{b} \bullet ad = \frac{1}{b} \bullet bc$	Multiply both sides of the equation by the same number $(\frac{1}{b})$.
$\frac{ad}{b} = c$	Multiplication.
$\frac{1}{d} \bullet \frac{ad}{b} = \frac{1}{d} \bullet c$	Multiply both sides of the equation by the same number $(\frac{1}{d})$.
$\frac{a}{b} = \frac{c}{d}$	Multiplication.

Facilitation

Encourage students to work in teams to figure out their own strategy for dealing with the challenging numbers. Observe, though, to ensure no one is getting too frustrated. If students seem unable to proceed, ask questions to help them understand their own reasoning.

Making the Lesson Easier

If students are struggling, you may want to use cards with whole numbers between 10 and 100 at first. You may also want to model your reasoning through several examples; then suggest that student pairs work together and think out loud about their rationalizations for rounding.

LESSON 8 IN ACTION

Helping students use their sense of multiplicative reasoning to estimate appropriately sometimes requires using smaller numbers first.

I thought this lesson might be challenging for students, but I never expected it to be so challenging for me as a teacher. I had been working with a core group of students through many of the lessons in this unit, so I felt they were pretty well grounded in proportional reasoning.

I began with three numbers: 271, 76, and 181. I placed them into a pair of ratios like so:

$$\frac{271}{76} = \frac{181}{?}$$

I then modeled my thought process by talking out loud. "I know that 27 and 18 are both multiples of 9, so 270 (271 rounded) and 180 (181 rounded) would be multiples of 90." Without much guidance they were able to see that 270 and 180 are the same ratio as 3 to 2.

I now rewrote the problem:

$$\frac{3}{76} = \frac{2}{?}$$

Students found this challenging. They had had exposure to the cross-product property, so they were willing to try to find the exact missing number. They thought that was much easier than trying to estimate for the missing number.

After we struggled through the example above, I tried comparing what I thought were less difficult numbers, beginning with numbers less than 100. Even then it was a struggle. For example, when I used the numbers 34 and 72, students felt they should round to 30 and 70. When I asked what half of 70 was, they could tell me it was 35; for them, though, it felt wrong not to change both numbers to the same whole unit. The idea of rounding based on "the digit to the right being 5 or less (or more)" was so ingrained in them that they balked at considering rounding one number in a pair up and the other one down.

We worked on simpler numbers and had quite a discussion about the need to do more than simply rely on a calculator for the answer. In the end, they agreed that without knowing approximately what the answer should be, they would be completely dependent on trusting their calculator (or pencil-and-paper calculation).

Donna Curry
Gardiner Adult Education Program, Gardiner, Maine

9
FACILITATING LESSON

The Indian Ocean Earthquake and Tsunami

What can your money buy?

Synopsis

This lesson provides a context for the number skills used in the previous lesson. Students use proportional reasoning to **convert** international **currency**. Students will work in the context of the December 16, 2004 tsunami and the time period shortly afterward. The world community, including the U.S., donated billions of dollars to aid the victims. But what did U.S. dollars buy in terms of Sri Lankan rupees, Thai bahts, and Indonesian rupiahs? Students use proportional reasoning and the tools they have learned to figure out conversions between currencies. When the numbers get increasingly more complicated, they use estimation before calculating more exact answers.

1. The whole class compiles what they collectively know about the Indian Ocean earthquake and tsunami, and individuals complete a cloze exercise.
2. Student pairs refer to a currency conversion table to rewrite stories, converting currency given in U.S. dollars to Sri Lankan currencies and vice versa.
3. Small groups choose a country to which to donate $500 (U.S.), select items to buy, and convert U.S. dollars to the currency of that country, first by estimating and then by finding more exact answers.
4. The class summarizes by listing examples for which they might use these skills.

Objectives

- Apply proportional reasoning concepts to international currency conversion
- Use estimation to predict for difficult numbers in proportion problems
- Determine the exact (or nearly exact) answer for a missing number in a proportion problem

Materials/Prep

- Calculators
- World atlas
- Internet access

Heads Up!

The currency conversion chart on p. 112 in the *Student Book* is based on January 2005 exchange rates. Currency rates vary day to day and especially year to year. Be aware that all examples in this lesson are based on exchange rates from January 2005, shortly after the tsunami hit. If you wish, look up current exchange rates on a web site such as http://www.oanda.com/convert/classic.

Opening Discussion

Begin by asking:

Have you heard of the Indian Ocean earthquake and tsunami that occurred on December 26, 2004? What do you know about it?

Find out whether anyone had relatives or friends in the countries affected. As students volunteer information, record what they say on the board.

Say:

Today's lesson is set in the United States in the time right after the event. Your task will be to translate U.S. currency to the currencies of the countries affected by the earthquake and tsunami. What did our donations buy for the victims? You can use reasoning with proportion to convert money from different countries.

Begin by completing the story on page 110 of your *Student Book* to see how many numbers you can figure out.

Activity 1: Tsunami Numbers in the News

Refer students to *Activity 1: Tsunami Numbers in the News (Student Book*, p. 110). After individuals have tried filling in the given numbers, have groups of three or four students check their level of agreement. Groups finishing early could examine the world map to mark where they think the earthquake's epicenter occurred, given what they already know.

Activity 2: Tsunami Survivor Aid in Sri Lanka

In the aftermath of the tsunami, many personal stories appeared on the Internet. Two of them are quoted in this activity.

Read aloud the first story. Refer to the currency conversion chart on *Student Book*, p. 112. Ask students to round these numbers to more manageable numbers as they begin to work with them.

Ask:

In early 2005, about how many Sri Lankan rupees equaled one U.S. dollar? (100 is a close estimate.)

Then ask pairs to rewrite the story in Sri Lankan rupees. First confirm that everyone agrees on the estimated answers, and then ask for more exact answers.

Read aloud the second story. Say:

This story is written in Sri Lankan rupees. What are approximate amounts in U.S. dollars?

After you estimate the answer, use a calculator to arrive at more exact amounts.

Ask students to generalize conversion methods (Problems 3–5). Then bring the class together to share methods.

Listen for various strategies. Did anyone use a table, set up a proportion, or multiply or divide?

Activity 3: What Money Can Buy

Post a list of five of the countries affected by this natural disaster: Thailand, Sri Lanka, Indonesia, India, and the Maldives.

Form groups and ask each group to choose a different country from among the five. Make sure all five are chosen.

Before they choose a country, ask students to look at the currency conversion table (*Student Book*, p. 112) and think about how they would round each of the currencies to make the numbers more manageable. They should see, for example, that there are about 60 Indian rupees to every U.S. dollar.

Each group should read the list prepared by Oxfam America and decide on the goods they would like to purchase for $500 U.S. After groups draw up a roster of items, each item should be converted to the currency of the country.

Groups take turns reporting to the class on what their money will purchase in the currency of their chosen country.

Summary Discussion

Refer students to *Reflections* (*Student Book*, pp. 141–144) and suggest that students list steps to remember how to convert currency.

Practice

Converting to U.S. Dollars, p. 116
Students find an international currency converter on the Internet and convert several foreign bills into U.S. dollars.

The Cost of a Big Mac®, p. 117
Students compare the price of a Big Mac in six countries. Teachers can extend this practice and incorporate some digital literacy by having students evaluate and download data from https://bigmacpricing.com/. This site compares local prices of Big Macs throughout the world to guide people on how to fairly price products and services based on different economic backgrounds. The data, sourced from https://github.com/TheEconomist/big-mac-data, can be copied and pasted into a spreadsheet to aid students in comparing items of interest to them, such as date, country, currency, and price.

At the Airport, p. 118
Offers practice thinking about exchange rates among three countries.

Extension

The Richter Scale and Earthquakes, p. 119
Gives students an opportunity to understand better the system for measuring the strength of earthquakes.

Your Own Country, p. 119
Students have a chance to share their own experiences with currency conversions.

Test Practice

Test Practice, p. 120

Looking Closely

Observe whether students are able to

Apply proportional reasoning concepts to international currency conversion

Currency conversion requires keeping track of the direction of the conversion. For example, given that 98.1195 rupees is equivalent to $1 U.S. in 2004, do students have a sense of whether to multiply or divide 500 rupees by 98.1195 to arrive at the number of dollars? What tools do they use to think about this? Are they able to set up a table of numbers or create a pair of equal ratios in proper order? Do they use the cross-product property to solve or check their answers? If necessary, work with students to come to some agreement about the numbers in general; for example, in rupees, the amount will sound larger. This will help students focus as they check the reasonableness of their answers.

Use estimation to predict for difficult numbers in proportion problems

Estimation is an essential skill when converting currency. Are students using estimation to predict and verify more exact calculations? Do they see that 98.115 rupees are approximately the same as 100 rupees? Do they easily multiply by powers of 10 (10, 100, 1,000)? If students seem to be struggling, you may want to provide some mental math practice to brush up on multiplying by powers of 10.

Determine the exact (or nearly exact) answer for a missing number in a proportion problem

What strategies do students use to find the exact or almost exact missing amount in a proportion? If students seem stumped, remind them of the tools they have studied and used. They may choose to use the cross-product property, or they may divide both ratios and compare the resulting quotients.

Are they able to use the calculator effectively? If some students have difficulty using the calculator, pair those students with others who are more adept. Do not assume that students can readily handle a calculator, especially when working with more complicated numbers involving decimals. Students may not realize that "9.00" can be entered into a calculator simply as "9." The fewer key strokes students make, the less likely they are to make careless errors.

WHAT TO LOOK FOR IN *LESSON 9*	WHO STANDS OUT? (LIST STUDENTS' INITIALS)			NOTES FOR NEXT STEPS
	STRONG	ADEQUATE	NEEDS WORK	
Concept Development • Reasons about monetary conversions • Rounds numbers based on a close relationship of the numbers • Predicts the missing amount based on a reasonable estimate				
Expressive Capacity • Explains the method used to convert currency				
Use of Tools • Effectively uses proportional reasoning tools learned in previous lessons				
Background Knowledge • Connects money conversion issues to personal experience				

Rationale

This lesson provides a context for the new cross-product tool explored in *Lesson 8* and focuses on a social studies context for applying proportional reasoning to problems with more complex numbers.

Math Background

No new math concepts are introduced in this lesson. Rather, students are encouraged to use skills developed throughout *Lessons 1–8*. However, the numbers in this lesson get much more challenging. Attention to detail is very important here. Currency conversions are often difficult; students will have to think carefully about the direction to take when converting from one currency to another.

Context

When this lesson was developed, the December 2004 tsunami was in the news. Much information was available on the Internet. Informative websites were

https://www.unicefusa.org/mission/emergencies/tsunamis/2004-south-asia

https://reliefweb.int/

https://www.oxfamamerica.org/

Some of the links will require students to apply filters and identify key words to search for material related to the December 26, 2004, tsunami. This provides an opportunity to work on important digital literacy skills; however, it may require additional instructor support. Use this site to guide students with online search techniques: https://edu.gcfglobal.org/en/search-better-2018/.

This might also be a good opportunity to talk about your students' countries of origin. They may appreciate the chance to explain the currency used in their native countries. If students have had to deal with currency conversions, they probably have developed their own strategies. Promote sharing.

Facilitation

Encourage students to work in teams to figure out their own strategies for dealing with challenging numbers. Make sure no one is getting too frustrated. If students seem to struggle, ask questions to help them understand their own reasoning.

Making the Lesson Easier

If the numbers are too difficult, the whole class could work together rounding the exchange rates to approximate whole numbers.

Making the Lesson Harder

Students may want to compare the exchange rates for today with those of January 2005. Has the tsunami disaster impacted the economy of the countries involved?

FACILITATING LESSON 10

As If It Were 100

Why are percents so commonly used?

Synopsis

This lesson starts out focusing on why and how percents are powerful tools for making relative comparisons. The percent ratio allows comparisons of unequal data sets by considering the base of comparison as 100. Percents level the playing field. For example, when trying to figure out whether 9 women out of 10 people is a relatively more female population than 42,000 women out of 45,000 people, it helps to ask: What if the total population in each situation were 100?

The contexts are demographics of education programs similar to your class and of the world population (where students use a base of 1,000). The numbers used are not very friendly, so students will need to use what they know about rounding and estimating.

1. The whole class examines local data about your educational institution. Either you will provide program data or the class will collect class data. Students then examine a set of comparable national data.
2. Pairs of students create ratios to describe each set of data and then compare the ratios in percent form (as if the base were 100).
3. Utilizing the world population data and the concept of the world as a village with 1,000 people, student pairs create equivalent ratios, using 100 as a base, and present a report on their findings.
4. The class summarizes why percents (and numbers such as 1,000 or 1,000,000) are useful when making comparisons.

Objectives

- Use percent as a way to make comparisons between data sets of different sizes, some with very large numbers
- Use 1,000 as a base for comparison between data sets of different sizes

Materials/Prep

- Data from your education program or class (if there are 10 or more students)
- Internet access
- Markers
- Easel paper

Prepare the following table for *Activity 2: What If There Were 1,000?*

Category	If There Were 1,000	If There Were 100	Actual Numbers

Heads Up!

The national statistics displayed in *Activity 1* are based on demographics of federally funded U.S. adult basic education programs. If you are in another type of program (e.g., a high school or a community college), you might want to find national statistics for your own type of educational program, such as the total number enrolled, gender, age ranges, and ethnicity.

Opening Discussion

Start by creating a set of data from your educational program or class.

Display the following, supplying the actual numbers.

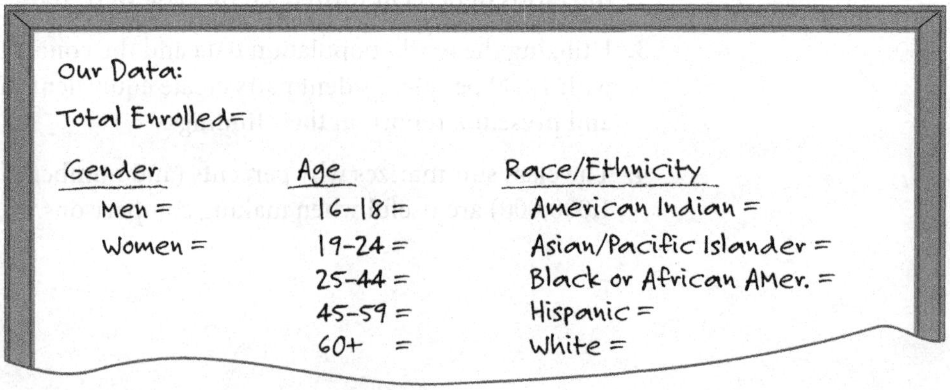

Ask:

What statements can you make about the makeup of our program (or class)?

How can you use percents to describe the gender breakdown?

There are several ways to do this, but encourage students to think of proportions. For example, if the total enrollment is 80 and there are 32 men, write:

$$\frac{32 \text{ men}}{80 \text{ total}} = \frac{? \text{ men}}{100 \text{ total}}$$

Ask:

Approximately what number of men would keep the ratios equal?

How could you get an exact answer?

Today you will compare the makeup of our program with national statistics. It will be helpful to use percents when making the comparisons.

Activity 1: How Does Your Program Compare?

Start by asking students to write the data about their program from the Opening Discussion onto their tables in *Student Book*, p. 123.

Begin by looking at the gender category. Ask:

How do our gender statistics compare with the national statistics?

How can you use percents to make the comparisons?

After agreement has been reached as to the relative gender makeup of the local and national populations, form small groups of students. Ask half the groups to begin with age and the other half to begin with ethnicity. However, each group should continue on with the other category after finishing the one assigned to them first.

The groups' task is to determine the breakdown of each category by percents, first for their program (or class) and then for the national data.

After completing the table, each group will prepare a report about the comparisons, including two circle graphs that illustrate the percent breakdowns.

Because the national numbers are very large, encourage students to round and estimate before they make exact calculations.

As groups report, probe for methods used.

After groups report, summarize:

Why are percents used for comparison?

Activity 2: What If There Were 1,000?

This activity has two purposes. It offers students the opportunity to grapple with very large numbers (in the billions) and to use 1,000 as a base of comparison in addition to 100(%).

Start by asking students to examine the information in *What if There Were 1,000?* (*Student Book*, p. 124).

Ask:

What strikes you about the information in the table?

Following some discussion, individuals should select one piece of information (such as the number of Christians) most interesting to them and then complete the story. After completing the story, suggest that pairs check each other's work.

When everyone is finished, call the group back together and call attention to the table you have prepared (see *Materials/Prep*). Have the class share their findings for each category.

Ask:

What patterns do you notice in the table?

What surprised you about the findings?

If time, consider showing this YouTube video "If The World Were 100 People" https://youtu.be/QFrqTFRy-LU to see how these patterns and findings would look if comparing similar categories for only 100 people.

Summary Discussion

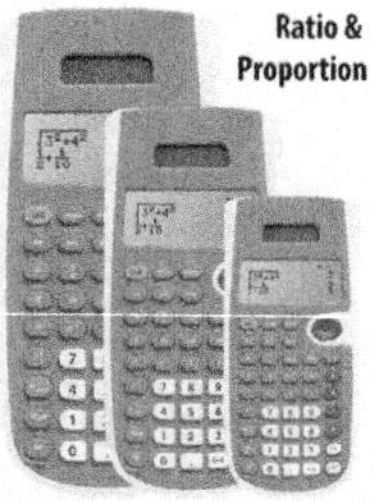

Summarize by asking a student to explain the value of using percents to make comparisons. Then ask students to respond in the *Reflections* section of the *Student Book* (pp. 141–144).

Practice

Percent Estimates, p. 126
Provides practice estimating what percent one number is of another.

Percent Dilemmas, p. 127
Gives students an opportunity to clarify a dilemma related to percents.

Free Throws—Who Is the Best? p. 128
Provides practice comparing absolute numbers and percents.

The Savings Contest, p. 129
Provides practice comparing absolute numbers and percents in a different context.

Pet Store Survey, p. 130
Offers students an opportunity to take raw data and develop it into a meaningful report using percents.

Comparing Your Program to the Community Adult Education Program, p. 131
Offers further practice making comparisons of unequally-sized data sets using percents.

Extension

Parts per Million, p. 132
Gives students a chance to move from the concept "out of 100" to "out of 1,000,000" in a work-related context.

Test Practice

Test Practice, p. 133

Looking Closely

Observe whether students are able to

Use percent as a way to make comparisons between data sets of different sizes, some with very large numbers

By the time students begin this lesson, they should have had multiple experiences with benchmark percents. They should be able to move quickly from fractions representing half to 50% and should be able to use a calculator to estimate other benchmark percents.

Even though this lesson includes some challenging numbers, the expectation is that students will round them in order to arrive at fairly close estimates.

Do you see evidence of students using various strategies: diagrams, the rule of equal fractions, unit rate, and the cross-product property? Students should now have a bank of tools for proportional reasoning. If any students seem to be getting bogged down using one strategy or tool, encourage them to consider what other tools they know how to use. Because you want students both to estimate and also to determine a more accurate answer, they may want to use one tool for estimation and another to arrive at a solution.

Use 1,000 as a base for comparison between data sets of different sizes

When the numbers being compared are very large (such as those representing world population), it is useful to use 1,000 as a base number, rather than 100. Encourage students to think about situations in which 1,000 is more useful than 100 for making comparisons. Students may not realize that 100 as a base comparison often leads to unwieldy decimals, and if they round, they lose some of the sense of the data. With 1,000 as a base, there is less rounding, therefore more exact comparisons.

In *Extension: Parts per Million*, students see yet another base for comparison: 1,000,000. Even if you do not assign the extension, you may want to be sure that students understand the value of using 100, 1,000, and 1,000,000 in making comparisons of unequally sized data sets.

WHAT TO LOOK FOR IN *LESSON 10*	WHO STANDS OUT? (LIST STUDENTS' INITIALS)			NOTES FOR NEXT STEPS
	STRONG	ADEQUATE	NEEDS WORK	
Concept Development • Understands the part:whole relationship necessary for working with percents • Can make comparisons of two unequal data sets using percents				
Expressive Capacity • Describes the value of percent in making comparisons • Can explain what percent means in terms of proportional reasoning				
Use of Tools • Uses appropriate tools to determine estimates and exact answers				
Background Knowledge • Builds on prior knowledge and experience using percents • Builds on knowledge of proportional reasoning				

Rationale

Percents are one of the most common ways of comparing data sets of different sizes. However, the concept of percent is not always clearly understood. This lesson focuses specifically on percent to illustrate its usefulness in and connection to proportional reasoning.

Earlier in this unit, students worked with benchmark percents. This lesson moves them toward complicated numbers and therefore "messy" percents. Students now have the tools and strategies to workl with these less friendly percents.

Math Background

We often think of percent as simply another way to write a fraction. The power of percent is revealed when comparing two data sets of different magnitude. It puts the data, no matter how disparate, on a level playing field. It allows for multiplicative comparisons between unlike sets of data.

Context

The data used in Activity 1 comes from statistics published by the National Reporting System (NRS) for Adult Education located at https://nrs.ed.gov/rt. Selecting the Reports tab and then Aggregate Reports will allow you to search for the most recent data by region or for all regions. Students also can research and evaluate data for their own state using the Reporting Tables tab.

Facilitation

If students are having difficulty, suggest that they first try to round to friendly benchmark percents and then decide whether the actual answer is more or less than the benchmark percent.

Making the Lesson Easier

For students who are struggling with the challenging numbers, first have them round all the numbers to friendlier numbers, perhaps to the nearest thousand, ten-thousand, or hundred-thousand, depending on the numbers and students' needs. It might also help to offer a new chart where students can record their rounded numbers so that they are not distracted by the unwieldy numbers.

Facilitating Closing the Unit: Reasoning with Ratios

Where do you find rates and ratios outside the math clasroom?

Synopsis

Throughout the unit, students have learned a variety of tools to aid proportional reasoning. They began by drawing pictures, looking at the multiplicative relationships across and within ratios (often using the property of equal fractions), and creating graphs. As the numbers got harder, they used additional tools to handle unwieldy numbers, so they were encouraged to compare ratios using unit rates and the cross-product property. This session is designed so students can show how well they can use the various tools to address each situation presented.

1. Students locate and analyze ratios in everyday life and create a final project report.
2. Students solve a set of open-ended tasks that call upon proportional reasoning.
3. Students create a Mind Map and compare it with the Mind Map from *Opening the Unit.*

Objectives

- Use various tools to address different situations involving proportionality
- Use ratios to make comparisons, predictions, and decisions

Materials/Prep

- Calculators
- Easel paper and markers for the *Final Project* poster
- Small packets of M&M's, one per student for the *Final Assessment*

Prepare one *Rubric For Final Project Presentations*, p. 139, per pair of students.

Prepare one *Final Assessment*, pp.159–163, one per student.

Prepare one copy of the *Final Assessment Class Tally*, p. 164.

Prepare one *Final Assessment Checklist*, p. 165, per student.

Opening Discussion

The final project, the *Final Assessment*, and the Mind Map give students the opportunity to show what they have learned in this unit. Explain to the class that they are going to use the different tools they have learned to solve several problems. By the end of this lesson, you will have three artifacts to assess what students have learned:

- The *Final Project: Rates and Ratios In Your Life*
- The *Final Assessment*
- A Mind Map

You might want to begin by reviewing what students remember, lesson by lesson. Use the table of contents to prompt students to recall what they have learned.

Ask:

What was your favorite lesson?

What was easiest for you? What was hardest?

Is there anything you wish to review before you attempt the final project and assessment?

Activity 1: Rates and Ratios In Your Life

Decide whether you wish to have students complete this final project individually or as pairs. It will be an opportunity for students to make yet another connection between what they are learning in class and how they use that learning in their lives.

Refer students to *Activity 1* in the *Student Book* (p. 136). Tell students to read the suggested list of places where they can investigate rates and ratios, and take other suggestions. Ask them to decide on a place to conduct their investigation. Point out that there are six questions they are to answer. They will need to take the assignment home and return to class with reports on what they find. Provide all students with the *Rubric for Final Project Presentations* (p. 139), so they know they will be assessed on the completeness of their reports. Before they leave, ensure that everyone understands the assignment.

You want to plan for a session where students present their final reports. In addition to addressing the prompts in a written report, students should be encouraged to prepare a poster, slideshow, or other digital presentation. As students present their reports, leave time for questions and comments from the rest of the class. You might want to ask the class to evaluate how well their peer presenters did by filling out the *Rubric for Final Project Presentations*.

If there is time, ask students to solve each other's word problems.

Activity 2: Final Assessment

There are four tasks in the *Final Assessment*. You might want to read them aloud together and ask students to first check off whether they feel they can do, cannot do, or are not sure if they can do the problems. Then ask students to complete first the tasks they feel most confident they can do, but also attempt the ones they are less sure they can do.

Activity 3: Mind Map

Ask students to fill in a Mind Map (*Student Book*, p. 137). Begin by asking who remembers filling in a Mind Map on the first day of this unit. Choose a volunteer to explain the mapping process. If no one recalls the process, provide a summary and an example of a Mind Map.

Summary Discussion

Ask students how they felt about the *Final Assessment*:

What did you feel particularly good about?

Were there any particular parts that you still feel unsure about?

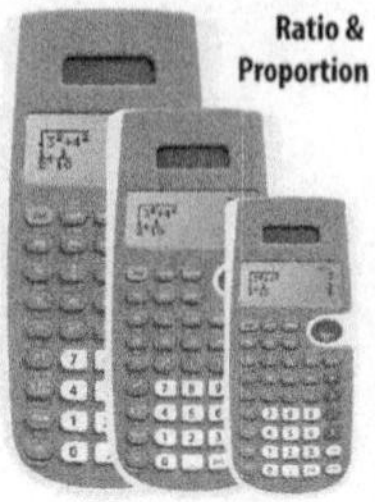

To close, ask students to share important things they have learned from working through this unit, first writing in *Reflections*, pp. 141–144, and then sharing their reflections.

Looking Closely

Observe whether students are able to

Use various tools to address different situations involving proportionality

Pay particular attention to the tools and their proper use. Are students able to use a second reasoning method to verify their answers? Do they use estimation to arrive at reasonable answers?

Use ratios to make comparisons, predictions, and decisions

How well are students able to navigate the open-ended investigation of the *Final Project*? Are they able to detect rates and ratios and to use the rates and ratios they found to predict and project into other amounts?

Are they able to do this well in the tasks in the *Final Assessment*, particularly the Statue of Liberty and M&M's tasks, where the numbers get more complicated?

RUBRIC FOR FINAL PROJECT PRESENTATIONS

Name(s)

____

____

____

____

____

This is an assessment of how well you addressed each of the questions in the *Final Project* assignment: *Rates and Ratios in Your Life.*

Give 0–4 points for each area.

0 = not at all
1 = somewhat, but vague
2 = adequate
3 = well done
4 = exceeds expectations

___ Clearly described where and what you investigated.

___ Listed several rates and ratios you discovered there. Included actual items you found or took a photo or drew a sketch.

___ Chose one of the ratios on which to focus. Predicted for at least two other amounts based on the ratio that you chose. Drew out your prediction and made a set of equal fractions to show how you made your prediction.

___ Showed the relationship with a graph.

___ Showed situations you investigated that were not proportional and described them, or imagined a related situation that would not be proportional.

___ Using the information you found, wrote a math word problem for someone else to solve.

___ Explained what was learned.

Notes

____

____

____

____

____

____

____

____

____

WHAT TO LOOK FOR IN *CLOSING THE UNIT*	WHO STANDS OUT? (LIST STUDENTS' INITIALS)			NOTES FOR NEXT STEPS
	STRONG	ADEQUATE	NEEDS WORK	
Concept Development • Demonstrates the underlying proportional relationships in everyday activities • Understands when situations are not proportional				
Expressive Capacity • Articulates how ratios are used in personal experiences				
Use of Tools • Uses tool effectively to solve proportion problems				
Background Knowledge • Uses personal experience to reflect on how ratios impact his or her life				

Rationale

This lesson presents several situations in which ratio and proportion are used to make predictions or decisions.

Math Background

Students should now be able to address situations involving proportional reasoning in a variety of ways. When the numbers are friendly, they may choose to use the multiplicative rule for fractions or may mentally look across or within numbers. They may even choose to create a table of equal ratios or draw a graph. When the numbers are messy, students should now have other tools to reference—e.g., the cross-product property or the unit rate (decimal quotient).

Facilitation

Because this lesson is an assessment, observe and listen as students work through the various situations.

Appendices

Name ______________________________ Date ______________

Keeping Things in Proportion: Reasoning with Ratios

INITIAL ASSESSMENT

First check off how you feel about your ability to solve each task. Then complete the tasks that you can do.

Look at each task and decide whether you can do it, don't know how to do it, or are not sure if you can. Check off the appropriate choice, and then complete the task if you can.

Task 1: Complete the Pattern

- Complete the pattern for each set of items below.
- There are no special deals for buying the larger amount.

1. 2 boxes for $8

    a. 6 boxes for ______

    b. ______ boxes for $20

___ Can do ___ Don't know how ___ Not sure

2. 4 pints for $12

    a. 12 pints for ______

    b. ______ pints for $24

___ Can do ___ Don't know how ___ Not sure

3. 5 lb. for $3

    a. 20 lb. for ______

    b. ______ lb. for $1.50

___ Can do ___ Don't know how ___ Not sure

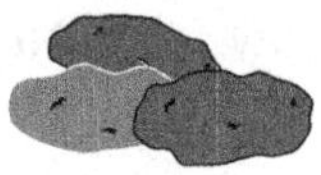

4. Describe how you figured out your answers.

Task 2: Making Predictions Based on a Survey

A survey was taken at a supermarket. Seven people said they preferred diet cola, and 28 said they preferred regular cola.

1. If you asked 100 shoppers, approximately how many do you expect would prefer diet cola?

___ Can do ___ Don't know how ___ Not sure

How did you figure out your answer?

2. If you asked 1,000 shoppers, how many do you expect would prefer regular cola?

___ Can do ___ Don't know how ___ Not sure

How did you figure out your answer?

Task 3: Which Is More Fully Grown?

A potted palm is 5 feet tall. When it is fully grown, it will be 10 feet tall. A fig tree is 6 feet tall. When it is fully grown, it will be 11 feet tall.

1. Which plant is closer to being fully grown?

___ Can do ___ Don't know how ___ Not sure

2. Explain why you think so.

Task 4: Signing Up

Akim was asked to create a jumbo-sized reproduction of a poster he had designed. His original design is 6″ wide and 18″ long.

1. If he keeps the same shape, how long would the cardboard be if a jumbo-sized sign is going to be 4 feet wide?

___ Can do ___ Don't know how ___ Not sure

2. If the original amount of cardboard cost $1, how much would you expect Akim to pay for the cardboard to make the jumbo reproduction?

___ Can do ___ Don't know how ___ Not sure

3. Use the grid below to show how the two poster sizes compare. Make the drawings to scale.

___ Can do ___ Don't know how ___ Not sure

Task 5: Test Practice

1. Which of the following graphs represents Aaron's wages if he earns $20 per hour?

A

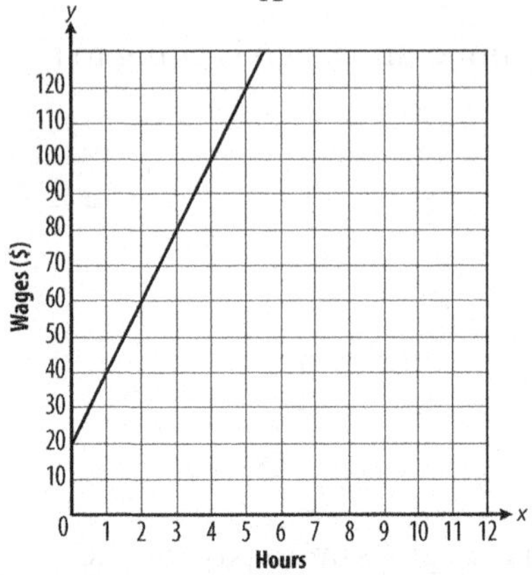

B

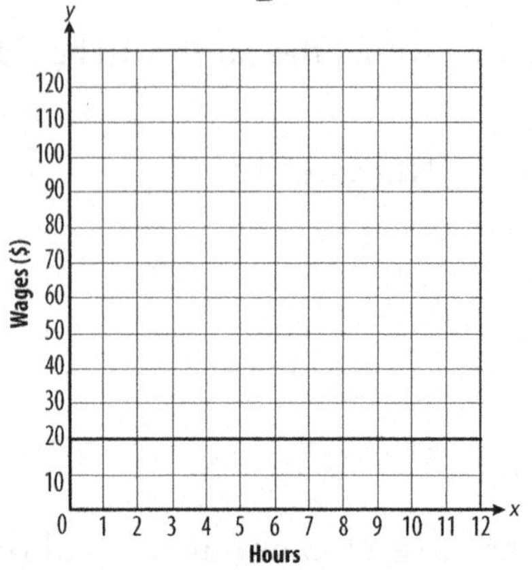

C

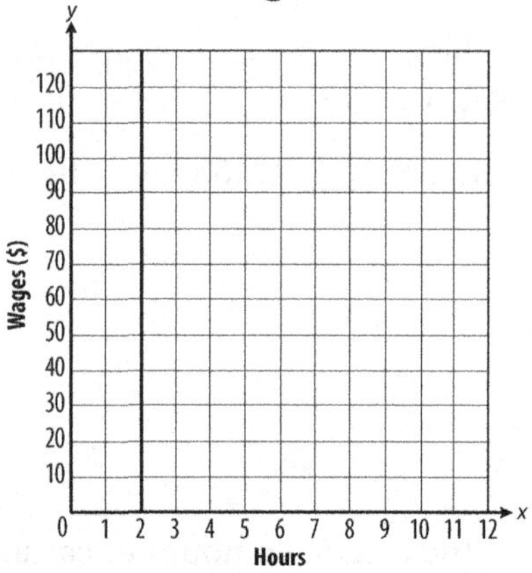

D

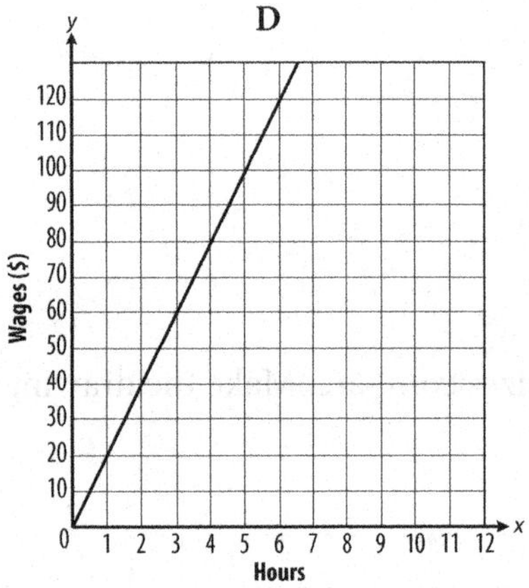

E

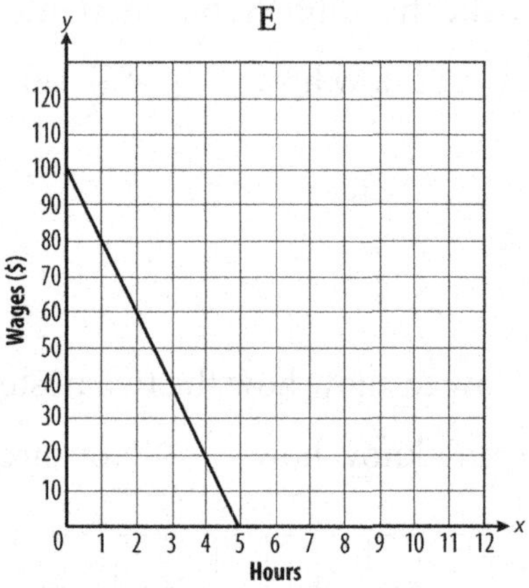

(1) A

(2) B

(3) C

(4) D

(5) E

___ Can do ___ Don't know how ___ Not sure

2. The graph below represents Jai's hourly wages. What does he earn per hour?

(1) $12.00

(2) $15.00

(3) $37.00

(4) $70.00

(5) Not enough information

___ Can do ___ Don't know how ___ Not sure

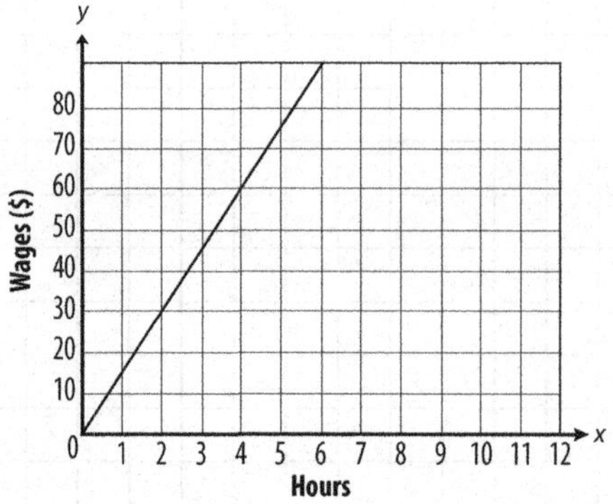

3. Maia walked 2 miles in 30 minutes. What could she do to figure how many miles it would take her to walk for one and a half hours if she is planning to keep the same pace?

(1) Add 3 miles to 2 miles

(2) Add 1.5 miles to 2 miles

(3) Multiply 2 miles by 3

(4) Multiply 2 miles by 1.5

(5) Subtract 30 minutes from one and a half hours

___ Can do ___ Don't know how ___ Not sure

4. A recipe calls for 2 cups of water to 1 cup of rice. Which of the adjusted recipes below would probably turn out too dry?

(1) 1 cup of water to $\frac{1}{4}$ cup rice

(2) 1 cup of water to $\frac{1}{2}$ cup of rice

(3) 4 cups of water to 1 cup of rice

(4) 4 cups of water to 2 cups of rice

(5) 4 cups of water to 3 cups of rice

___ Can do ___ Don't know how ___ Not sure

5. Tom picked a bathroom tile pattern that uses 16 black tiles for every 24 white tiles. If the floor requires 200 tiles, about how many black tiles would Tom need?

(1) 40

(2) 60

(3) 80

(4) 133

(5) 160

___ Can do ___ Don't know how ___ Not sure

6. What number would x represent to make the following proportion true?

$$\frac{50}{40} = \frac{400}{x}$$

___ Can do ___ Don't know how ___ Not sure

INITIAL ASSESSMENT CLASS TALLY

Task	Can Do	Don't Know How	Not Sure	Percent Who Can Do
1. 1a				
1b				
2a				
2b				
3a				
3b				
2. 1				
2				
3. 1				
2				
4. 1				
2				
3				
5. 1				
2				
3				
4				
5				
6				

Notes on Confidence Levels

INITIAL ASSESSMENT CHECKLIST

Use a ✓, ✓+, or ✓– to assess how well students met each skill. When you give feedback to students, note areas in which they did well in addition to areas for improvement.

Use ✓ to show work that is mostly accurate; some details, additional work needed

Use ✓+ to show work that is accurate, complete.

Use ✓– to show work that is inaccurate, incomplete.

Student's Name______________________________

Task	Skills	Solution Strategy (Tools)					
		Diagram	Scale Up/Down	Unit Rate	Graph	Cross Product	Other
1. Complete the Pattern 1 ____ 2 ____ 3 ____	Given the cost for some items, calculates costs for other amounts (at the same rate)						
2. Survey 1 ____ 2 ____	Recognizes part/part/whole Predicts for 100 shoppers at the same rate Predicts for 1,000 shoppers at the same rate						
3. More Fully Grown 1 ____ 2 ____	Uses multiplicative reasoning Uses additive reasoning						
4. Signing Up 1 ____ 2 ____ 3 ____	Finds the missing length Considers area as contributing to cost Draws to scale						
5. Test Practice 1 ____ 2 ____ 3 ____ 4 ____ 5 ____ 6 ____	Interprets graph Identifies strategy Identifies quality Identifies part to whole Solves for x in proportion						

OVERALL NOTES

Strengths

Areas for Improvement

Name ______________________ Date ______________

Keeping Things in Proportion: Reasoning with Ratios

INTERIM ASSESSMENT

Task 1: A Puzzling Situation

The pieces of the puzzle you have been given form a 4″ × 4″ square.

Work in teams of three.

- One team member should carefully cut out each of the six pieces for the rest of the team.
- Then each team member chooses two pieces.

Your team should make a new set of puzzle pieces so that when the new pieces are put together, they create a square that measures 10″ × 10″.

1. When put together, did the new pieces create a square?

2. How would you explain to someone else how you figured out the dimensions of the new pieces? Write the directions for creating the pieces for the new square.

3. What is the area of the original square?

4. Predict the area of the new square.

5. Measure the area of the new square. How does it compare with your prediction?

6. Explain with pictures or words the relationship between the area of the original square and the area of the new square.

Task 2: Recipe Project

1. Find a recipe. Write the ingredients in the column on the left in the following table. Then double the recipe. Next triple the original recipe. Then halve the original recipe. Then make enough for only one serving, and write the new amounts in the last column. (Sometimes you do not keep things exactly in proportion for recipes. Is there anything that you would be watchful for when changing proportions?)

Ingredients	Double the Original Amount	Triple the Original Amount	Half the Original Amount	Amount Needed for One Serving

2. Now choose any of the two ingredients in your recipe. Look at the ratio between these two ingredients in all versions. Show that the ratio remained the same with a visual representation and a fraction form.

3. Use the grid paper below to draw a graph to show that the two sets of ingredients stayed in balance.

Task 3: How Low Can You Go?

Today you plan to buy five grocery items. Here is your grocery list:

- 1 lb. of pasta
- A jar of pasta sauce (24-oz. size)
- Five dozen rolls of paper towels
- 10 lb. of potatoes
- 1.5 qt. container of ice cream

You are able to get to three supermarkets (All Foods Market, Stop 'n' Save, and Eddy's Foodtown).

1. At which supermarket will you buy each item if you plan to spend the least amount of money?

2. What is the lowest total cost of the five items if you can shop at all of the three supermarkets? Show your work.

Task 4: Test Practice

1. Which is *not* an equal ratio to 5:8?
   (1) 2.5:4
   (2) 10:16
   (3) 20:32
   (4) 24:15
   (5) 25:40

2. Which of the rates is equal to 2 pounds for $6?
   (1) 1 pound for $4
   (2) 3 pounds for $7
   (3) 4 pounds for $10
   (4) 5 pounds for $15
   (5) 6 pounds for $10

3. Which of the following graphs represents Shareen's hourly wage of $14 an hour?

A

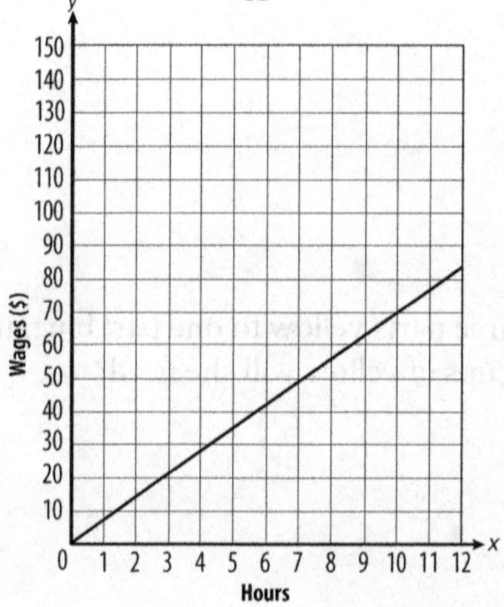

B

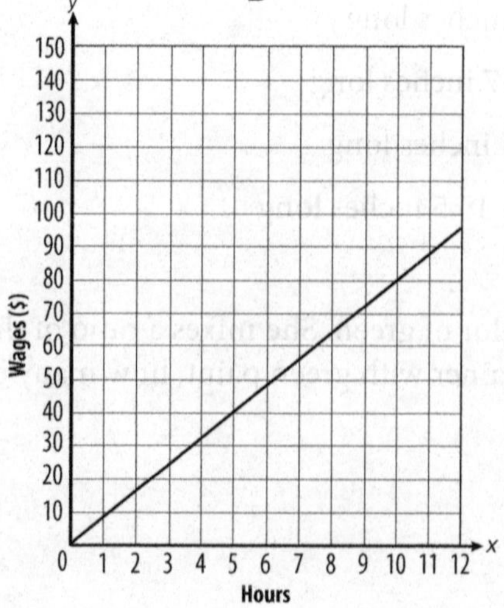

C

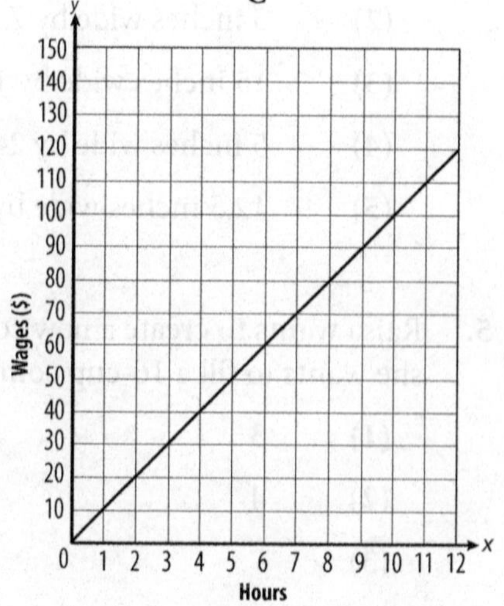

D

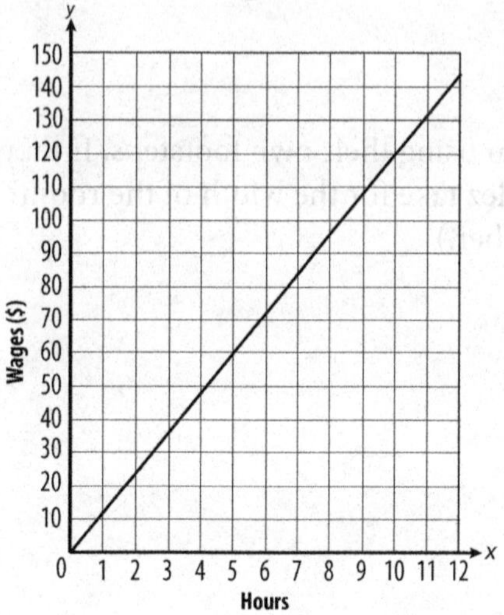

E

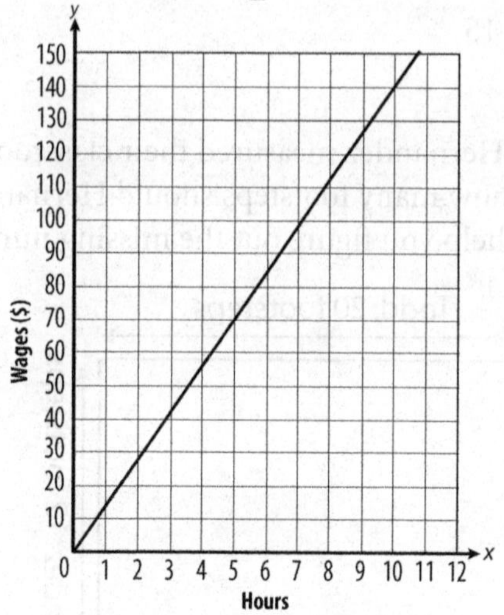

(1) A

(2) B

(3) C

(4) D

(5) E

4. Which of the following frames would be proportional to a frame that is 10 inches wide and 12 inches long?

(1) 5 inches wide by 6 inches long

(2) 5 inches wide by 7 inches long

(3) 15 inches wide by 17 inches long

(4) 5 inches wide by 24 inches long

(5) 12.5 inches wide by 15.5 inches long

5. Raisa wants to create a new color of green. She mixes a ratio of three parts yellow to one part blue. If she wants to fill a 16-cup container with green paint, how many cups of yellow will she need?

(1) 3

(2) 4

(3) 7

(4) 12

(5) 15

6. Todd and Hernandez measured their classroom using their own footsteps. If they kept their steps the same, how many footsteps should Hernandez take for the width of the room? (Use the following picture to help you figure out the missing number.)

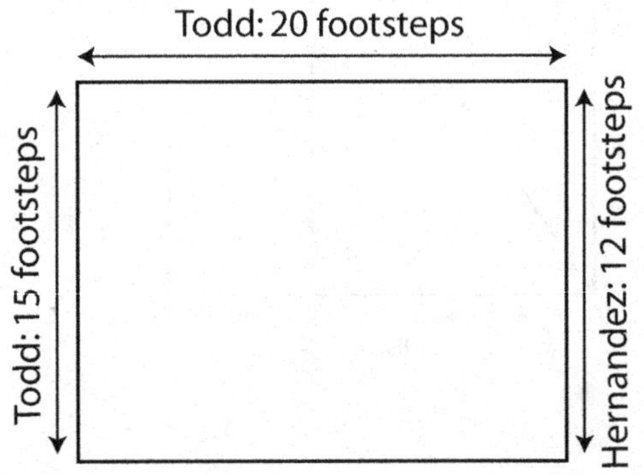

Name ______________________________ Date ______________

Keeping Things in Proportion: Reasoning with Ratios

FINAL ASSESSMENT

Task 1: Statue of Liberty: The Nose Knows

The Statue of Liberty is 111 feet 1 inch tall, from her heel to the top of her head. Her nose is 4 feet 6 inches long. Estimate how long the length of her right arm is. Show how you determined your answer.

___ Can do ___ Don't know how ___ Not sure.

Task 2: Sampling M&M's

You will need a small bag of M&M's for this task. Only eat them after you are done with the math!

1. Open up the bag and count the number of red M&M's inside it. Using that information, predict how many red M&M's you would expect to find in a one-pound bag of M&M's. Show your work here.

___ Can do ___ Don't know how ___ Not sure.

2. Do you predict that there would be the same number of brown M&M's as red ones in a one-pound bag? Why or why not?

___ Can do ___ Don't know how ___ Not sure.

Task 3: Making a Mixture

Maia mixed 2 ounces of lemon concentrate with 18 ounces of water to make a lemon mixture. Roberto mixed 3 ounces of lemon concentrate with 20 ounces of water to make his mixture. Whose mixture had the stronger lemon flavor? Explain how you figured out your answer.

___ Can do ___ Don't know how ___ Not sure.

Task 4: Traveling Buddies

Bick's car travels at a rate of 9 miles every 12 minutes. Ann's car travels at a rate of 45 miles an hour. If both cars start at the same time, whose car will reach point A, 10 miles away, first? Show your thinking.

___ Can do ___ Don't know how ___ Not sure.

Task 5: Test Practice

1. Derek built a frame that was 15 inches long and 10 inches wide. Then he decided to build one that was the same shape but *20* inches long. Which of the following proportion set-ups could he use to figure out the width of his frame?

   (1) $\frac{15}{10} = \frac{w}{20}$

   (2) $\frac{15}{10} = \frac{20}{w}$

   (3) $\frac{15}{20} = \frac{w}{10}$

   (4) $\frac{20}{10} = \frac{w}{15}$

   (5) $\frac{15}{w} = \frac{10}{20}$

___ Can do ___ Don't know how ___ Not sure

2. Xai had a recipe that called for 3 cups of flour and $\frac{1}{2}$ tsp. baking powder. She used $1\frac{1}{2}$ cups of flour and $\frac{1}{4}$ tsp. of baking powder. To keep the recipe in the right proportion, what does she need to do to the rest of the ingredients?

   (1) Double the amount of each of the ingredients.

   (2) Use half the amount of each of the ingredients.

   (3) Use one-quarter the amount of each of the ingredients.

   (4) Subtract $1\frac{1}{2}$ the amount of each of the ingredients.

   (5) Use one-third the amount of each of the ingredients.

___ Can do ___ Don't know how ___ Not sure

3. Which of the following is *not* a true proportion statement?

   (1) $\frac{4}{6} = \frac{2}{3}$

   (2) $\frac{5}{10} = \frac{20}{40}$

   (3) $\frac{5}{8} = \frac{2.5}{4}$

   (4) $\frac{7}{12} = \frac{24}{14}$

   (5) $\frac{8}{9} = \frac{4}{4.5}$

___ Can do ___ Don't know how ___ Not sure

4. On the straight road shown here, the distance between Ware and Penn is 45 miles. What is the distance between Litchfield and Penn?

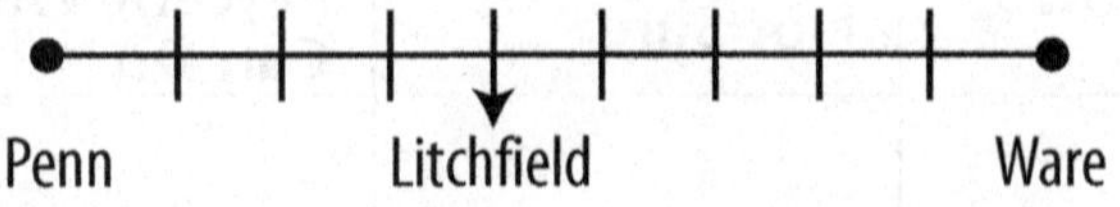

(1) 4

(2) 8

(3) 20

(4) 25

(5) 40

___ Can do ___ Don't know how ___ Not sure

5. At any time of the day when the sun is out, objects cast shadows that are in proportion to their heights. If a meter stick casts a 2.5 meter shadow, how tall would you expect a flag pole that casts a 40 meter shadow to be?

(1) 10 m

(2) 16 m

(3) 40 m

(4) 82 m

(5) 100 m

___ Can do ___ Don't know how ___ Not sure

6. In May 2005, the currency exchange rate between Australia and the United States was: 1 US dollar (USD) = 1.32091 Australian dollars (AUD). Approximately how many US dollars were $1,000 AUD worth?

___ Can do ___ Don't know how ___ Not sure

FINAL ASSESSMENT CLASS TALLY

Task	Can Do	Don't Know How	Not Sure	Percent Who Can Do
1.				
2. 1				
2				
3.				
4.				
5. 1				
2				
3				
4				
5				
6				

Notes on Confidence Levels

FINAL ASSESSMENT CHECKLIST

Use a ✓, ✓+, or ✓– to assess how well students met each skill. When you give feedback to students, note areas in which they did well in addition to areas for improvement.

Use ✓ to show work that is mostly accurate; some details, additional work needed.

Use ✓+ to show work that is accurate, complete.

Use ✓– to show work that is inaccurate, incomplete.

Student's Name____

Task	Skills	Solution Strategy (Tools)					
		Diagram	Scale Up/ Down	Unit Rate	Graph	Cross Product	Other
1. Statue of Liberty	Given information about dimensions, uses own body as guide to determine new dimensions						
2. M&Ms 1 ____ 2 ____	Determines the part/whole relationship Uses smaller package weight to determine the number of red M&Ms in a one-pound package						
3. Making a Mixture	Accurately compares the part/whole relationships Clearly explains reasoning						
4. Traveling Buddies	Accurately compares rates						
5. Test Practice 1 ____ 2 ____ 3 ____ 4 ____ 5 ____ 6 ____	Chooses correct proportion set up Identifies strategy Identifies incorrect proportion set-up Uses scale correctly Applies proportion to shadow situation Calculates currency conversion						

OVERALL NOTES

Strengths

Areas for Improvement

All Foods Market

2/99¢ with card

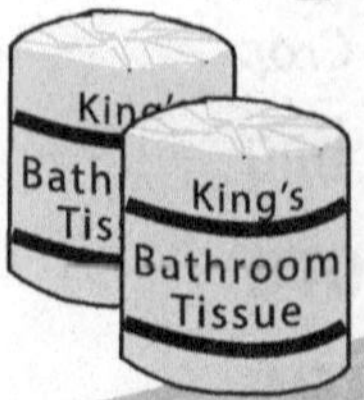

King's Single Roll
Bathroom Tissue
1000 sheets per roll

Sparklette Paper Towels

6-Roll Pack
360 Sq. ft.

$5 49

Basta Pasta

16 oz. All Varieties

5/$4

Paolo's Pasta Sauce

Assorted Varieties 24 oz.

2 for $4

1.5 Qt All Varieties

Dairyland Ice Cream & Frozen Yogurt

3/$8

US #1 "New Crop"

5 Lb. Bag Idaho Potatoes

$3 69 ea.

Check out the savings in every aisle.

Eddy's FOODTOWN

"The best prices in town."

16 Oz. Elbow Macaroni, Vermicelli, Regular or Thin Spaghetti

Pasta di Toscana Spaghetti

2/$1

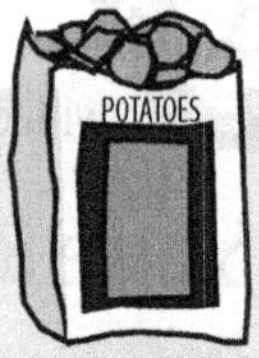

New Crop

10 Lb. Bag

Eastern Shore Potatoes

$5.79 ea.

Classics Pasta Sauce

24 oz.

2/$3

Bountiful Quilted Paper Towels

60 Sheets Single Roll

98¢

Emmy's Ice Cream

1.5 quart Select Varieties

$1.89

Price with $10 Purchase
Limit 2 pkgs.

SavMor 20 Roll Bath Tissue

$15.98

Stop 'n Save

ABSORBO
12 Pack
Paper Towels
10.99

Farm Fresh
Maine
Russet Potatoes
5 Lb. Bag

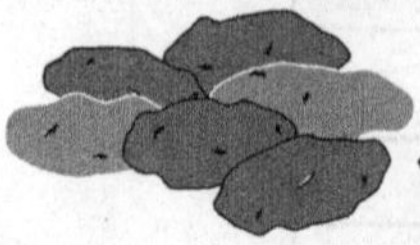

2 for $6

Paesano
Pasta Sauce
Assorted Varieties. 24 Oz.

5 for $8

Buy 2 – Get 1 FREE!

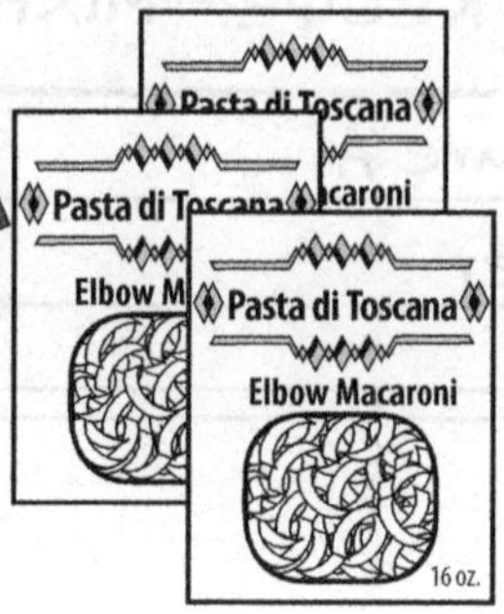

Pasta di Toscana
16 oz.
$1.09
(Excludes Gluten-free)

FROSTY
Ice Cream
Assorted Flavors
1.5 Qt
2 for $5

Fluffy Cloud
Bathroom Tissue
Single Ply— 12-Pack
2 for $12

Extra Savings

Blackline Master 4: Orange Juice Recipe

Orange Juice Recipe—Mixture A
3 cans of water
1 can of orange juice concentrate

Orange Juice Recipe—Mixture B
1 can of water
1 can of orange juice concentrate

Orange Juice Recipe—Mixture C
16 fl. oz. Mixture A
16 fl. oz. water

Blackline Master 5: Alternate Orange Juice Recipe

Orange Juice Recipe—Mixture A
1 can of water
1/3 can of orange juice concentrate

Orange Juice Recipe—Mixture B
1/3 can of water
1/3 can of orange juice concentrate

Orange Juice Recipe—Mixture C
2 1/3 cans of water
1/3 can of orange juice concentrate

a. In a survey of 100 people, Harvest Blend Bread was preferred to Sunshine Sliced Bread, 4:1.

b. In taste tests, about 80% of the people polled preferred Harvest Blend Bread to Sunshine Sliced Bread.

c. In taste tests, 5,975 people preferred Harvest Blend Bread and only 1,486 preferred Sunshine Sliced Bread.

d. In taste tests, about four out of five people we asked preferred Harvest Blend Bread to Sunshine Sliced Bread.

Blackline Master 7: Mona Lisa Cards A–C

A

B

C

Blackline Master 8: Mona Lisa Card D

D

Blackline Master 9: Mona Lisa Card E

E

F

G

Blackline Master 12: Pancake Recipe

PANCAKES

INGREDIENTS

1 1/4 cups of sifted flour
1/2 teaspoon of baking soda
1 egg, beaten
2 tablespoons of melted butter
1 tablespoon of baking powder
4 teaspoons of sugar
1 1/2 cup of buttermilk

Blackline Master 13: A Puzzling Situation

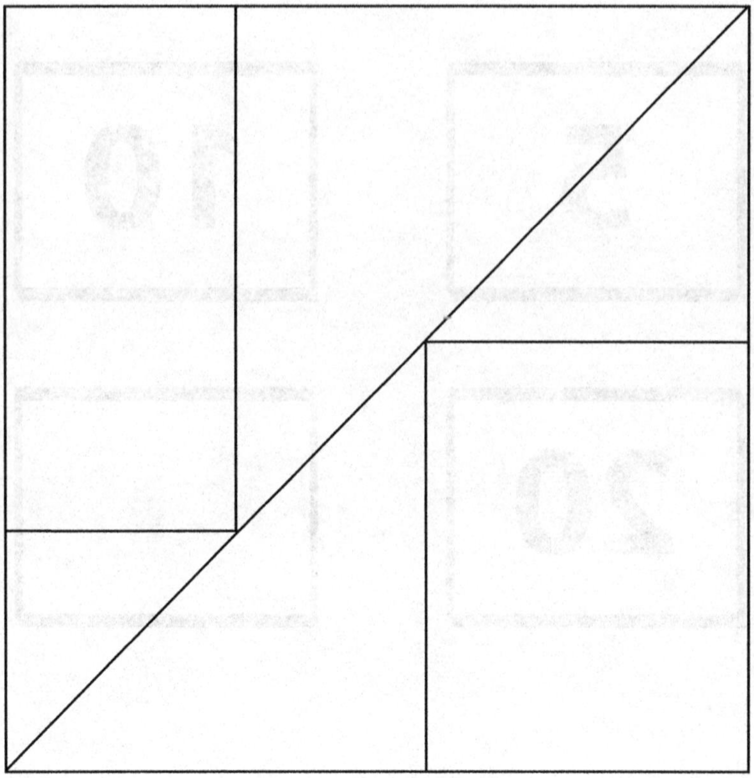

Blackline Master 14: Card Set 1—Friendly Numbers

5	10
20	

Blackline Master 15: Card Set 2–More Challenging Numbers

251	76
123	

Blackline Master 16: 0.5 cm Grid Paper

Blackline Master 17: 0.5 Inch Grid Paper

Blackline Master 18: 0.25 Inch Grid Paper

Glossary

Reminder: Students generate their own definitions for terms as they arise in class, using language that makes sense to them. However, to help you guide the discussion, we include mathematical definitions for most of the terms.

Italicized terms are *not* included in the *Student Book* but are included in the *Teacher Book* as background information. The lesson number in which the term first appears is in parentheses following the term.

area (6)—the number of square units needed to cover a surface. For example, the area of a 2″ x 11″ rectangle is 22 sq. in.

averaging **(2)**—computing or finding an average. An average is a numerical value used to describe the overall clustering of data in a set. Average is usually used to indicate an arithmetic mean. However, an average can also describe the middle of the data set (median) or the most typical value(mode).

comparison (Opening)—the act or process of examining two amounts in order to discover the relationship between them. Numbers may be compared either additively or multiplicatively. For example, one could compare two numbers by finding the difference between them. For example, $400 could be compared with $50 by saying that $400 is $350 more than $50, or $50 is $350 less than $400; those would be additive comparisons. On the other hand, one could compare $400 with $50 by focusing on the multiplicative relationship between the two amounts, saying that $400 is eight times as great as $50, or $50 is one-eighth of $400; those would be multiplicative comparions

concentrate (3)—a food substance, especially a liquid, made thicker or stronger in flavor by the removal of liquid. Frozen orange juice concentrate is an example.

convert (9)—change units of one measurement system into units of another. For example, one can convert U.S. dollars to Euros.

cross-product property (8)—for two ratios, if $a/b = c/d$, then $ad = bc$. For example, since 3/4 = 9/12, then 3 x 12 = 4 x 9 (note both equal 36).

currency (9)—a system of money, or the bills and coins themselves, used in a particular country.

dimensions (6)—the measure of the magnitude, or size, of an object. For example, the dimensions of a rectangle are its length and width. The dimensions of a cube are the length, width, and height. Their values describe how large the object is.

direct proportion **(Introduction)**—the relation between two quantities whose ratio remains constant.

fraction form (1)—the written expression of a ratio or quotient with a "/". For example, the fraction form of the ratio of 3 to 4 is 3/4.

graphing **(5)**—the act of creating a graph. A set of equal ratios may be shown by graphing on a coordinate plane the relationship between the two quantities compared. The graph will result in a straight line which passes through the origin (0, 0).

mixture (3)—a substance contining two or more ingredients combined or blended together. For example, an orange juice mixture can be a mixture of frozen orange juice concentrate and water.

part to part (4)—one meaning of a ratio that represents a relationship between one part of a whole and another part of the same whole.

part to whole (4)—one meaning of a ratio or fraction that represents a relationship between a part and its whole.

per (Opening)—for each or for every. The rate or ratio 60 miles/hr can be expressed as 60 miles per hour.

percent (4)—a fraction or ratio in which the denominator is 100. The reatio of a number to 100. Perccent means "per hundred." For example, 25% is 25 out of 100.

perimeter (6)—the sum of the lengths of the sides of a polygon. For example, the perimeter of a 2″ x 11″ rectangle is 26 in. Its four sides measure 2 in., 2 in., 11 in., and 11 in.

predict (Opening)—to say what is going to happen in another case on the basis of given information.

property of equal fractions, or **property of equal ratios (1)**—the mathematical principle that $a/b = (a \times c)/(b \times c)$, where c is any non-zero number. For example, $3/4 = (3 \times 5)/(4 \times 5) = 15/20$. This is the application of the multiplicative identity principle, that is, $a/b \times 1 = (a/b) \times (c/c)$.

proportion (1)—an equation which states that two ratios are equal.

***proportionality* (Opening)**—the state of quantities remaining in the same (multiplicative) relationship, even when the amounts change.

rate (1)—a type of ratio, used in many everyday problems, such as grocery shopping, traveling, and medicine—in fact, almost every activity involves some type of rate. Miles per hour or feet per second are both rates of speed. Number of heartbeats per minute is called "heart rate." If you ask a babysitter, "What is your rate?", you are asking how many dollars per hour you will be charged. The little word "per" is a clue that you are dealing with a rate. Unit price is a particular rate that compare s a price to some unit of measure.

ratio (1)—a multiplicative comparison of two amountsnumbers. When we express ratios in words, we use the word "to"—we say "the ratio of something to something else." Ratios can be written in several ways: as a fraction (3/4), with the word "to" (3 to 4), or with a colon (3:4).

sample (2)—a selection from a population. Usually we cannot collect data from every member of the population in which we are interested but must find a sample that will accurately represent that population. We use the results of the sample to draw a conclusion about the population. Here are two different types of samples:

biased sample—a sample that does not accurately represent the population. Biased samples can give inaccurate results. For example, if you wanted to know the percentage of students at your school who recycle, sampling from the ecology club would result in a biased sample.

random sample—a sample that accurately represents the whole population. For example, if you wanted to get a random sample of all the students in your school, you could put all of their names in a bin and randomly draw out as many as you wanted in your sample.

similar (6)—similar figures have the same shape but may not have the same size. The measures of corresponding sides in similar figures are multiples of each other (i.e., they are proportional).

***similarity* (5)**—the quality or state of being similar, that is, not differing in shape, but only in size or position.

speed (7)—the rate of motion, measured in distance/time. For example: the speed at which your car travels may be 55 miles per hour.

surface area (6)—area of any enclosed two-dimensional space. Often, the sum of the areas of all the faces, or surfaces, of a solid figure.

tsunami (9)—a great sea wave produced especially by submarine earth movement or volcanic eruption.

unit rate (7)—a comparison of two measurements in which the second term has a value of 1. Speed limit is a unit rate because it is measured as the number of miles traveled per 1 hour. So, 55 miles per hour is a unit rate, whereas 110 miles per 2 hours is not a unit rate (because the second measurement, time, has a value of 2, not 1).

Answer Key

Opening the Unit

Activity 1: What Would It Take?

Answers will vary. Sample dream job:
I would like having a telecommuting job so I could live half of the year in Arizona during the winter and spring months and in New England during the summer and fall. I would need to cover costs for housing for both locations, as well as computer equipment and travel every six months. I estimate I would need $100,000 a year.

Activity 2: Initial Assessment

Task 1: Complete the Pattern

1. **a.** 6 boxes for $24
   b. 5 boxes for $20
2. **a.** 12 pints for $36
   b. 8 pints for $24
3. **a.** 20 lb. for $12
   b. 2 1/2 lb. for $1.50
4. Explanations will vary.

Task 2: Making Predictions Based on a Survey

1. About 20; reasons will vary. Sample reason: The total number of people surveyed was 35. There are just over 3 sets of 35 in 100, so 3 x 7 = 21, which is close to 20.
2. About 800; reasons will vary. Sample reason: In 100 people, there would be about 80 who prefer regular cola. In 1,000, there would be about 10 times as many, or 800.

Task 3: Which Is More Fully Grown?

1. Answers will vary. Sample answer using additive reasoning: Both plants are equally close to being grown.
   Sample answer using percents: The fig tree is closer to being fully grown.
2. Answers will vary. Sample answer using additive reasoning: 10 – 5 = 5; 11 – 6 = 5, so both plants will grow another 5 feet before being fully grown.
   Sample answer using percents: The potted palm is 50% of its height. The fig tree is about 55% grown, so only has about 45% of its growth left.

Task 4: Signing Up

1. 12′
2. $64
3. 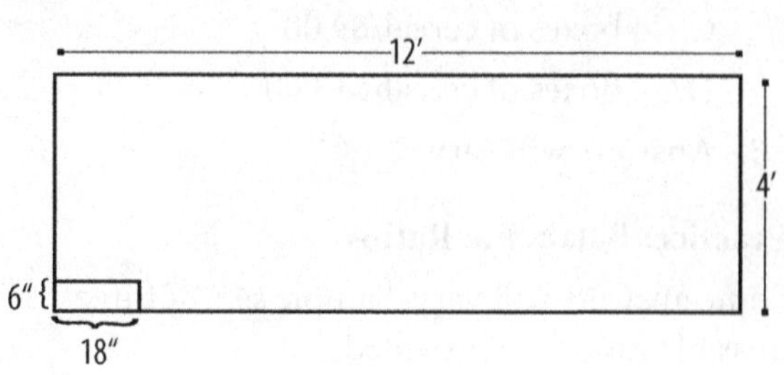

Task 5: Test Practice

1. (4)
2. (1)
3. (3)
4. (5)
5. (3)
6. 320

Activity 3: Mind Map

Answers will vary.

Lesson 1: A Close Look at Supermarket Ads

Activity 1: Different Numbers—Same Ratio

Answers will vary. Sample response:
ad: canned vegetables, 10 cans for $5.00.

$$\frac{10\text{ cans}}{\$5.00} = \frac{5\text{ cans}}{\$2.50} = \frac{2\text{ cans}}{\$1.00} = \frac{8\text{ cans}}{\$4.00}$$

Activity 2: From Pictures to Mathematical Rules

Answers will vary. Sample responses based on sample ad above:

1.

$$\frac{10 \text{ cans}}{\$5.00} = \frac{5 \text{ cans}}{\$2.50} = \frac{2 \text{ cans}}{\$1.00} = \frac{8 \text{ cans}}{\$4.00}$$

2. To get from 10 cans for $5.00 to 5 cans for $2.50, I divided both numbers by 2.
3. When I halve the top number, I also halve the bottom number, so both numbers are growing or shrinking at the same rate.
4. In the drawing, the numbers also got bigger or smaller at the same rate.

Activity 3: Buying in Quantity

1. Answers will vary.
2. Answers will vary. Sample response based on the following ad: 2 boxes of cereal for $5.00.
   A. 2 boxes of cereal/$5.00
   B. 6 boxes of cereal/$15.00
   C. 4 boxes of cereal/$9.00
   D. 5 boxes of cereal/$13.00
3. Answers will vary.

Practice: Balancing Ratios

Some answers will vary. Sample sets of three possible ratios are provided.

1.

$$\frac{\$1.00}{1 \text{ bag}} = \frac{\$4.00}{4 \text{ bags}} = \frac{\$10.00}{10 \text{ bags}}$$

2.

$$\frac{2 \text{ bottles}}{\$6.00} = \frac{1 \text{ bottle}}{\$3.00} = \frac{4 \text{ bottles}}{\$12.00}$$

3.

$$\frac{1 \text{ lb.}}{\$0.80} = \frac{5 \text{ lbs.}}{\$4.00} = \frac{10 \text{ lbs.}}{\$8.00}$$

4.

$$\frac{\$3.50}{4 \text{ muffins}} = \frac{\$7.00}{8 \text{ muffins}} = \frac{\$10.50}{12 \text{ muffins}}$$

5.

$$\frac{1 \text{ 12-pack}}{\$6.00} = \frac{3 \text{ 12-packs}}{\$18.00} = \frac{4 \text{ 12-packs}}{\$24.00}$$

6.

$$\frac{\$2.00}{4 \text{ cartons}} = \frac{\$4.00}{8 \text{ cartons}} = \frac{\$6.00}{12 \text{ cartons}}$$

Practice: Using the Property of Equal Fractions

1. Multiply the number of pounds by 3 and multiply the cost by 3.
2. Multiply the number of cartons by 15 and multiply the cost by 15.
3. Halve the number of bags and halve the cost.
4. Multiply the number of pizzas by 10 and multiply the cost by 10.
5. Answers will vary.

Practice: Is Bigger Cheaper?

1. Answers will vary. Sample answer: No, the cost of the larger one with twice as much detergent is the same as the cost of 2 of the smaller ones. 1 box/$4.00 = 2 boxes/$8.00
2. It has more than twice the amount of detergent as the smaller box.

Practice: Working with Larger Numbers

Some answers will vary. Sample answers included.

1.

$$\frac{2 \text{ lbs.}}{\$15.00} = \frac{100 \text{ lbs.}}{\$750.00} = \frac{250 \text{ lbs.}}{\$1{,}875.00}$$

2.

$$\frac{2 \text{ lbs.}}{\$1.50} = \frac{300 \text{ lbs.}}{\$225.00} = \frac{10 \text{ lbs.}}{\$7.50}$$

3.

$$\frac{12 \text{ rolls}}{\$3.50} = \frac{240 \text{ rolls}}{\$70.00} = \frac{48 \text{ rolls}}{\$14.00}$$

4.

$$\frac{12 \text{ rolls}}{\$3.50} = \frac{600 \text{ rolls}}{\$175.00} = \frac{24 \text{ rolls}}{\$7.00}$$

Practice: Mental Math

1. 2/3
2. 8,000/10,000
3. Divide each amount by 100.
4. Divide each amount by 5.
5. Multiply each amount by 60.
6. Divide each amount by 500.

7. Divide each amount by 100.
8. Multiply each amount by 2,000.

Extension: Ratios in My Life

Answers will vary.

Test Practice

1. (4)
2. (1)
3. (2)
4. (1)
5. (4)
6. $150.00

Lesson 2: It's a Lot of Work!

Activity: How Long Will It Take?

Answers will vary. Look at *Lesson 2 in Action* for a sample response.

Practice: Time to Be Proportional

1. 72 hours
2. 3 hours
3. 48 hours/2days = 72 hours/3 days
4. 120 seconds/2 minutes = 3,600 seconds/60 minutes
5. 2 minutes/120 seconds = 4 minutes/240 seconds
6. 12 hours/0.5 days = 48 hours/2 days

Practice: Wasting Water

Answers will vary.

Practice: Typing Tests

1. January: 12.5 minutes;
   February: 10 minutes;
   March: 10 minutes;
   April: 5 minutes
2. Answers will vary. Sample answer: In January, Kayla's typing speed was 20 words per minute, which meant she took 12.5 minutes to type a page of 250 words. By February, she had increased her speed to 25 words per minute. This meant she took 10 minutes to type a page. She held the same speed in March, but by April, she really increased her typing speed. In April, she was typing 50 words per minute, double what she did in February and March. It took her only 5 minutes to type a page.
3. Answers will vary. Sample answer:
   February: 125/5 = 250/10 = 25/1
   March: 200/8 = 100/4 = 25/1

Practice: Who Is the Fastest?

1. Drake: 20 miles per hour; Leon: 22.5 miles per hour; Denton: 25 miles per hour
2. Answers will vary. Sample answer:
   Drake = 50 miles/2.5 hours = 100 miles/5 hours; Denton = 100 miles/4 hours (which is not the same as 100 miles/5 hours for Drake); Leon = 45 miles/2 hours = 90 miles/4 hours (which is not the same as 100 miles/4 hours for Denton).

Test Practice

1. (1)
2. (3)
3. (4)
4. (3)
5. (3)
6. 5.5

Lesson 3: Tasty Ratios

Activity 1: Orange Juice Taste Test

Answers will vary. Look at *Lesson 3 in Action* for a sample response.

Activity 2: "Doctoring" the Orange Juice

Answers will vary for the third column. Sample answers:

Mixture A 3:1

Mixture B 1:1

Add two more parts water to the mixture. That will make 3 parts water to 1 part mixture.

Mixture C 7:1

Add two more parts water to make the mixture 9:1. Then add two more parts concentrate, which will now make the mixture 9:3, which is the same ratio as 3:1.

Practice: How Does It Taste?

1. Too salty
2. Just right
3. Too sour
4. Too dry
5. Too vinegary

Practice: Lemonade for the Party

1. a. 4 1/3 cans
   b. 8 servings, since 1 can makes 2 qt. (64 fl. oz.) lemonade
   c. 52 fl. oz. water to 12 fl. oz. lemonade concentrate, or 4 1/3 parts to 1 part concentrate
2. a. 2 cans
   b. 8 2/3 cans (104 fl. oz.)
   c. 128 fl. oz = 4 qt. = 1 gal.
3. Answers will vary. Sample answer: Put a little less water into the original mixture.

Practice: Building a Set of Equal Ratios

Answers will vary. Sample answers:

1. 2/5 = 4/10 = 20/50 = 40/100 = 60/150
2. 1/4 = 5/20 = 25/100 = 100/400 = 200/800
3. 3/4 = 6/8 = 12/16 = 60/80 = 90/120
4. 4/5 = 8/10 = 12/15 = 60/75 = 120/150

Practice: Popcorn Party Mix

1. Answers will vary for the third column. Sample answers:

Original Recipe	Triple the Recipe	Show How You Did It
2 qt. popcorn	6 qt. popcorn	2 x 3 = 6
1 c. pretzels	3 c. pretzels	1 x 3 = 3
1 1/2 c. cheese curls	4 1/2 c. cheese curls	3 x 1 = 3, 3 x 1/2 = 1 1/2 3 + 1 1/2 = 4 1/2
3/4 c. salted peanuts	2 1/4 c. salted peanuts	3/4 = 1/2 + 1/4 3 x 3/4 = 1 1/2 + 3/4 = 2 1/4
1/4 c. butter	3/4 c. butter	1/4 + 1/4 + 1/4 = 3/4
1 T. Worcestershire sauce	3 T. Worcestershire sauce	1 x 3 = 3
1/2 tsp. garlic salt	1 1/2 tsp. garlic salt	1/2 + 1/2 + 1/2 = 1 1/2

2. 7 1/2 quarts of party mix. Sample answer:
   2 1/2 x 3 = 2 x 3 + 1/2 x 3 = 6 + 1 1/2 = 7 1/2
3. Answers will vary. Sample answer:

$$\frac{\text{3/4 c. butter}}{\text{7 1/2 qt. party mix}} = \frac{\text{1/4 c. butter}}{\text{2 1/2 qt. party mix}}$$

because

1/4 c. butter—2 1/2 qt. mix.
1/4 c. butter—2 1/2 qt. mix.
1/4 c. butter—2 1/2 qt. mix.

Total: 3/4 c. butter—7 1/2 qt. mix.

Practice: Dressing It Up

1. a. 3 c. mayonnaise
   9 T. milk
   6 T. cider vinegar
   3 garlic cloves
   b. 1 c. mayonnaise:3 T. milk
   c. 3 c. mayonnaise:9 T. milk, or 1:3
2. a. 4 c. mayonnaise
   12 T. milk
   8 T. cider vinegar
   4 garlic cloves
   b. 3 T. milk:2 T. cider vinegar
   c. 12 T. milk:8 T. cider vinegar, or 3:2

Practice: Doctor This

Answers will vary. Sample answers:

1. a. He could add one more egg.
   b. He will have to double the remaining ingredients.
   c. 2
2. a. He could cut the recipe in half.
   b. He could halve all the ingredients.
3. a. He could use 2/3 of the amount of flour; 2/3 of 6 cups = 4 cups.
   b. She will have to take 2/3 of each of the ingredients.
   c. 2/3 of 5 dozen = 2/3 of 60, which is 40 (20 + 20 + 20 = 60, so 2/3 of 60 = 40).

Extension: Reasoning with Ratios

Explanations will vary. Sample explanations:

1. A; Pitcher A already has the strong taste; adding more syrup will only make it stronger. Adding water to Pitcher B will make it even weaker tasting.

2. B; Because there is less punch in Pitcher B to begin with, adding 1 c. of strawberry syrup will make a stronger tasting mixture than adding 1 c. syrup to Pitcher A.
3. A; Pitcher A is already stronger tasting. Adding the same amount of syrup to both pitchers makes them both stronger, but Pitcher A will still be the stronger of the two.
4. It depends on how strong each was before the syrup was added. Adding syrup to Pitcher A made it stronger tasting, while adding water to Pitcher B made it weaker tasting. There is not enough information to know.

Test Practice

1. (2)
2. (4)
3. (3)
4. (2)
5. (2)
6. 1

Lesson 4: Another Way to Say It

Activity 1: Part-to-Part versus Part-to-Whole

1. Answers will vary. Sample answers:

Water to Concentrate	3:1
Concentrate to Water	1:3
Concentrate to Mixture	1:4
Mixture to Concentrate	4:1
Mixture to Water	4:3

2. Water to concentrate; concentrate to water
3. Water to mixture; concentrate to mixture; mixture to concentrate; mixture to water

Activity 2: Two Truths and a Lie

1. Answers will vary. Sample answer:
   Truth—50% of Mixture B is water.
   Lie—7% of Mixture C is water.
2. Answers will vary.

Activity 3: Making Advertising Statements

1. Answers will vary. Sample answers:
   a. HHHH:S
   b. H (20%)
      H (20%)
      H (20%)
      H (20%)
      S (20%)
   c. 6,000/1,500 = 12,000/3,000 = 12/3 = 4/1
      HHHH/S
   d. HHHHS
2. All four send the same message.
3. Answers will vary.

Practice: What is Another Way to Say It?

1. Answers will vary. Sample answer:
   70% of people surveyed preferred Misty Mouthwash to Refresh Mouth Rinse.
2. Answers will vary.
3. Answers will vary.

Practice: Keeping the Ratio the Same with Gardeners

1. Answers will vary. Sample answer:
   9/6 = 3/2 = 9,000/6,000, 300/200.
2. 270
   [9 + 6 = 15; 9/15 = 270/450, because 15 x 30 = 450 and 9 x 30 = 270].

Practice: Who Is Right?

Both are right.
Sample answer to show Shaanti is right:
Total students who responded = 40.
Fifteen wanted Saturday classes, so 15/40 = 3/8.
Sample answer to show Bronson is right:
Fifteen students wanted Saturday classes to 25 students who did not. 15/25 = 3/5.

Practice: Rounding Ratios

1. 8,000 to 4,000
2. 7,000 to 8,000
3. 4,000 to 10,000
4. 7,000 to 9,000
5. 900 to 4,000
6. 9,000 to 15,000
7. 5,000 to 19,000

Test Practice

1. (1)
2. (3)
3. (5)
4. (4)
5. (2)
6. 60%

Lesson 5: Mona Lisa, Is That You?

Activity 1: How Large Is the *Mona Lisa* Painting?

1. 20 inches
2. Answers will vary.
3. Actual size should be 20 inches wide and 30 inches tall.

Activity 2: Mona Lisa, Is That You?

Part 1

1. Answers will vary.
2. Good reproductions:
   A—2″ x 3″
   D—5″ x 7 1/2″
   F—6″ x 9″
   C—3″ x 4 1/2″
   Bad reproductions:
   B —3″ x 2″
   E—6″ x 6″
   G—4.5″ x 9″

Part 2

1. $\frac{\text{width}}{\text{height}} \quad \frac{2}{3} = \frac{5}{7\frac{1}{2}} = \frac{6}{9} = \frac{3}{4\frac{1}{2}}$
2. Answers will vary. Sample answer: The ratio of width to height is always 2 to 3.
3. Answers will vary. Sample answer: 6/6 = 2/2 = 3/3; neither of these is the same as 2/3.
4. Answers will vary.

   Sample answer for 3″ x 2″: Change 2″ to 4 1/2″ for a ratio of $\frac{3}{4\frac{1}{2}}$.

   Sample answer for 6″ x 6″: Change 6″ to 4″ for a ratio of 4/6.

   Sample answer for 4.5″ x 9″: Change 4.5″ to 6″ for a ratio of 6 to 9.

Activity 3: Lining Up the Reproductions

See *Teacher Book, Activity 3*, for sample answers.

Practice: Rectangle Ratios

1. No
2. No
3. No
4. Yes
5. Yes
6. No
7. Yes
8. Height/width = 1/3 = 2/6 = 4/12 = 3/9

Practice: What Is the Pattern?

1. a. Multiply by 3.
   b. Divide by 2.
   c. Multiply by 4.
   d. Divide by 10.
   e. Divide by 6.
2. Answers will vary.

Practice: Using a Graph to Predict

1.

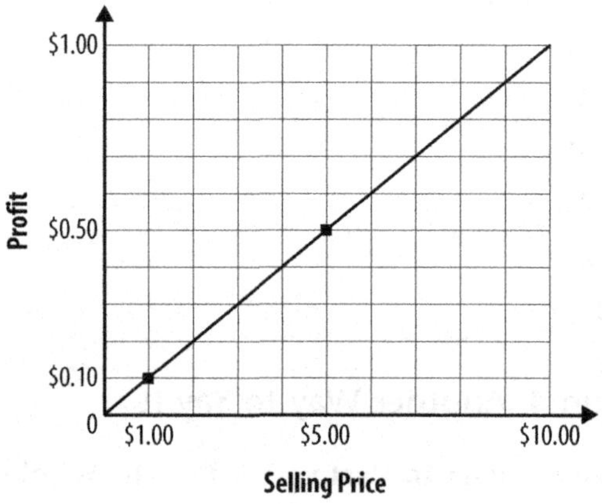

2. $10.00
3. c
4.

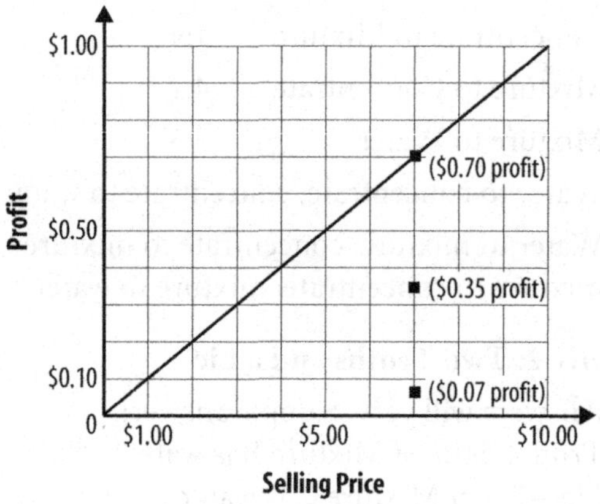

Practice: Graphing Ingredients

1. Answers will vary. Sample answers:
   a. 3 cups of water to 2 cups of rice
      No Too dry
   b. 5 cups of water to 2 1/2 cups of rice
      Yes Just right
   c. 6 cups of water to 3 cups of rice
      Yes Just right

d. 7 cups of water to 3 cups of rice
No Too watery

e. 9 cups of water to 4 1/2 cups of rice
Yes Just right

2.

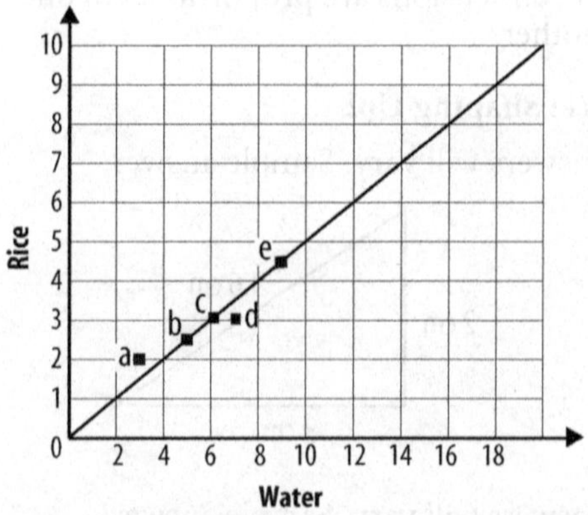

3. Answers will vary. The 2:1 ratios should all fall in line.

Extension: Paint by Numbers

Part 1

1. Good reproductions:
   18″ x 27″; 1 1/2′ x 2 1/4′

   Bad reproductions:
   8″ x 10″; 3″ x 5″; 9″ x 11″; 2′ x 4′
2. Answers will vary.

Part 2

Answers will vary. Sample answer: Those pictures that are distorted do not fall on the line. The pictures that measure 3″ x 5″ and 2′ x 4′ fall above the line. The pictures that measure 8″ x 10″ and 9″ x 11″ fall below the line. The picture that measures 18″ x 27″ (1 1/2′ x 2 1/4′) falls on the line.

Test Practice

1. (3)
2. (4)
3. (3)
4. (1)
5. (4)
6. 6″

Lesson 6: Redesigning Your Cell Phone

Activity 1: Three Cell Phones

1. Answers will vary. Measurements below are in centimeters and based on the Apple iPhone 11. Regardless of device type, the large-screen version's measurements will be double the original's. The compact version's dimensions will be half the original's, and one-fourth of the large-screen version's.

a.	7.6	15.2	3.8
b.	15.1	30.2	7.6
c.	1.0	2.0	0.5
d.	1.0	2.0	0.5
e.	6.7	13.4	3.4
f.	14.0	28.0	7.0

2. The large-screen dimensions are twice those of the original's. The compact-size dimensions are half the original's. The large-screen dimensions are four times the compact-size dimensions.
3. Answers will vary. Sample answer:
   15.1 / 30.2 = 1.0 / 2.0 = 6.7 / 13.4

Activity 2: Material for a Cell Phone Cover

1. Answers will vary.
2. **a.** 7.6 cm 15.2 cm 3.8 cm
   b. 15.1 cm 30.2 cm 7.6 cm
   c. Reasonable estimates: 105–120 sq. cm
   Reasonable estimates: 450–460 sq. cm
   Reasonable estimate: 24–32 sq. cm
   d. 114.8 sq.cm. 459.0 sq. cm 28.9 sq. cm
   e. 45.4 cm 90.8 cm 22.8 cm
3. No

Note: The sample answers for items #4, #6, and #7 below use rounded whole numbers to provide easier mental math calculations. Students needing more precise answers can calculate the area to the nearest hundredth before performing the appropriate operation.

4. Answers will vary. Sample answer:
   When the dimensions double, the area quadruples (e.g. using rounded area numbers, 115 x 4 = 460).
5. No
6. Answers will vary. Sample answer:
   When the dimensions are halved, the area is one-fourth that of the original (e.g. using rounded area numbers, 115 ÷ 4 ≈ 29).
7. Answers will vary. Sample answer:
   Large-screen version = about 460 sq. cm
   Compact version = about 29 sq. cm.
   You will need a little more than 15 times as much material for the large-screen version (or about 430 more sq. cm).

8. The perimeter doubles (45 cm x 2 = 90 cm).
9. The area quadruples (115 sq. cm x 4 = 460 sq. cm).

Practice: Similar Rectangles

1. A, H, and I; B, D, and F; C, E, and G
2. [All measurements in cm]
   A (3 x 5), H (6 x 10), and I (4.5 x 7.5); 3/5 = 6/10 (double 3 and 5) and 3/5 = 4.5/7.5 (multiply 3 and 5 by 1.5).
   B (4 x 4), D (6 x 6), and F (5 x 5); 4/4 = 5/5 = 6/6 because they are all 1/1 ratios.
   C (1 x 5), E (2 x 10), and G (1.5 x 7.5); 1/5 = 1.5/7.5 (multiply 1 and 5 by 1.5) and 1/5 = 2/10 (double 1 and 5).
3.

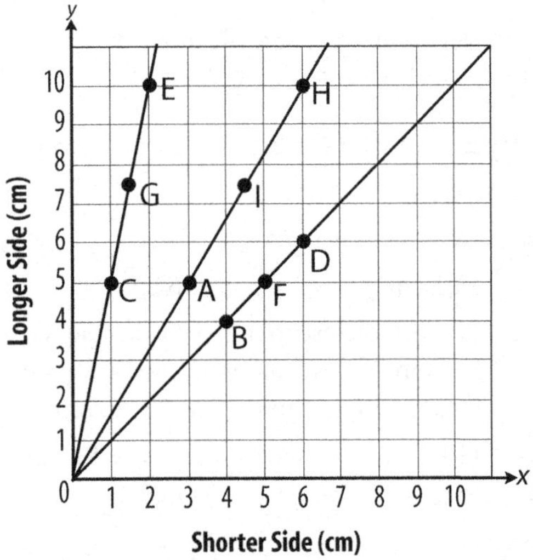

Answers will vary. Each group (A, H, and I; B, D, and F; C, E, and G) fall on a straight line

Practice: Graphing Calculator Information

1.

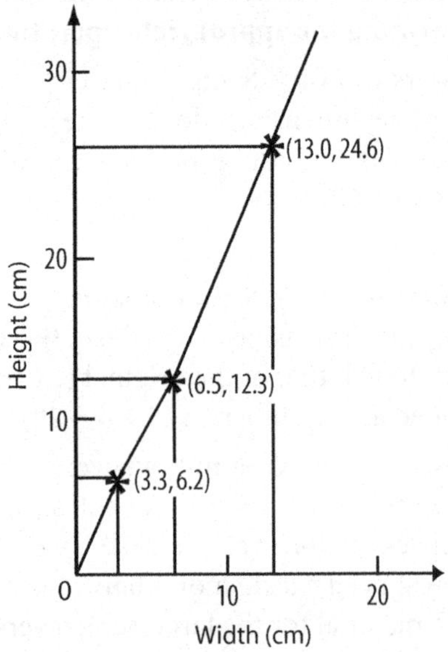

Answers will vary. Sample answer:
The end points lie along a straight line that begins at zero.

2. Answers will vary. Sample answer:
   The dimensions are proportional to one another.

Practice: Shaping Up!

1. Answers will vary. Sample answer:

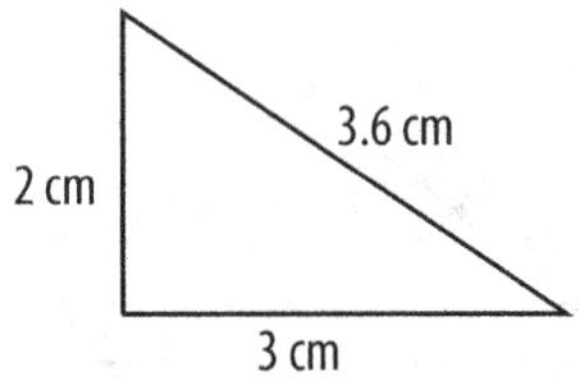

2. Answers will vary. Sample answer:

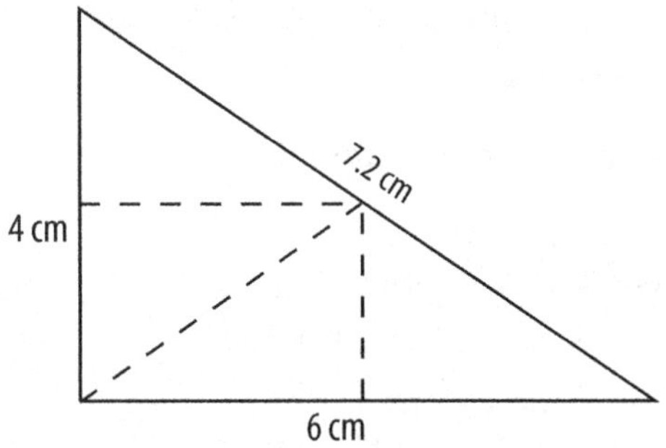

3. 4 (See dotted lines in triangle in 2.)
4. The angles stayed the same.

Practice: Squaring Off

1. 5 ft. to 10 ft., or 1 to 2
2. No. Sample explanation: The new area is four times as large as the original area.

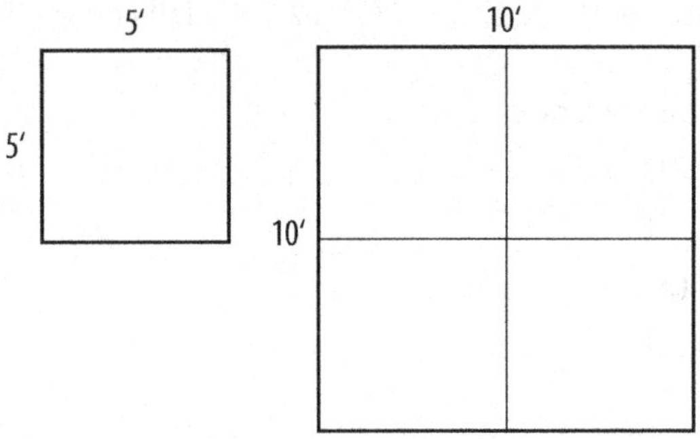

3. 100

4. Four. Sample explanation:

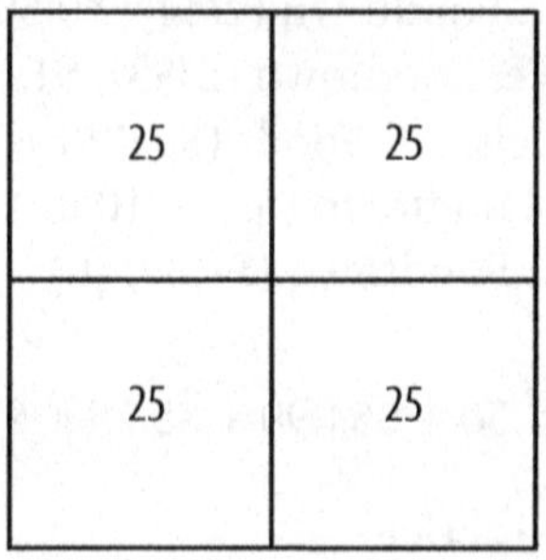

Extension: Perimeter, Surface Area, and Volume

1. a. 4 ft. 8 ft.
   b. 2 ft. 4 ft.
   c. 3 ft. 6 ft.
   d. 52 sq. ft Answers will vary. 208 sq. ft.
   e. 24 cu. ft. Answers will vary. 192 cu. ft.
2. The length, width, and height of the new box are double the original sizes.
3. The surface area became four times as large because the area of each side quadrupled. There will be four times as much material needed for the outside of the box.
4. When the dimensions doubled, the volume became eight times as large. There will be eight times as much space for packing.

Test Practice

1. (3)
2. (4)
3. (2)
4. (4)
5. (5)
6. 3

Interim Assessment

Task 1: A Puzzling Situation

1. Answers will vary (but should be yes).
2. Answers will vary. Sample explanation: Take one puzzle piece. Measure the edges of the piece. Multiply the length of each of the edges by 2.5. Make a new piece that has edges that measure 2.5 times the original edges, keeping the shape the same.
3. 16 sq. in.
4. Answers will vary. A good prediction is 100 sq. in.
5. Answers will vary.
6. Answers will vary. Sample explanations: The area of the new shape is 6.25 times larger than the original because each side is 2.5 times bigger (2.5 x 2.5 = 6.25). 16 sq. in. x 6.25 = 100 sq. in.

   Or, according to the shape below,

   4 x (4″ x 4″) = 64 sq. in.

   4 x (2″ x 4″) = 32 sq. in.

   1 x (2″ x 2″) = 4 sq. in.

   Total 100 sq. in.

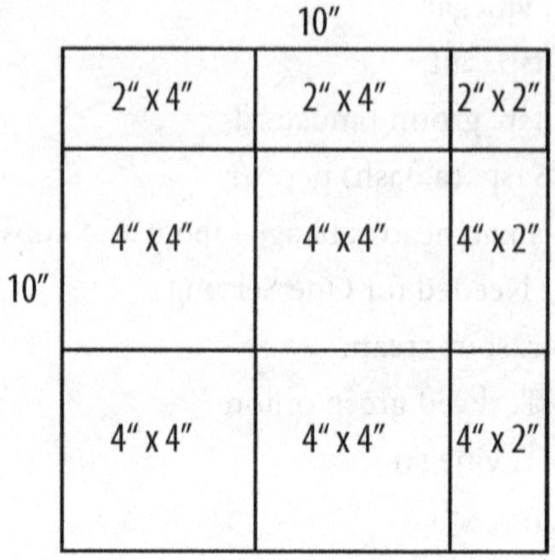

Task 2: Recipe Project

1. Answers will vary. Sample answer based on Sour Cream Cole Slaw Recipe, which serves 6:

   1 c. sour cream

   2 T. sliced green onion

   2 T. vinegar

   1 1/2 tsp. salt

   1/4 tsp. ground mustard

   1/8 tsp. pepper

   1 small head cabbage (about 7 cups)

   Double the Original Amount

   2 c. sour cream

   4 T. sliced green onion

   4 T. vinegar

   3 tsp. salt

   1/2 tsp. ground mustard

   1/4 tsp. pepper

   2 small heads cabbage (about 14 cups)

Triple the Original Amount

- 3 c. sour cream
- 6 T. sliced green onion
- 6 T. vinegar
- 4 1/2 tsp. salt
- 3/4 tsp. ground mustard
- 3/8 tsp. pepper
- 3 small heads cabbage or 2 large heads (about 21 cups)

Half the Original Amount

- 1/2 c. sour cream
- 1 T. sliced green onion
- 1 T. vinegar
- 3/4 tsp. salt
- 1/8 tsp. ground mustard
- 1/16 tsp. (a dash) pepper
- 1/2 small head cabbage (about 3–4 cups)

Amount Needed for One Serving

- 1/6 c. sour cream
- 1/3 T. sliced green onion
- 1/3 T. vinegar
- 1/4 tsp. salt
- dash of ground mustard
- dash of pepper
- 1/6 small head cabbage (about 1–1 1/2 cups)

2. Answers will vary. Sample answer based on sample above: sour cream/green onion

$$\frac{1\text{c.}}{2\text{ T.}} = \frac{2\text{c.}}{4\text{ T.}} = \frac{3\text{ c.}}{6\text{ T.}} = \frac{1/2\text{ c.}}{1\text{ T.}} = \frac{1/6\text{ c.}}{1/3\text{ T.}}$$

3. Answers will vary. Sample graph based on sample answer above:

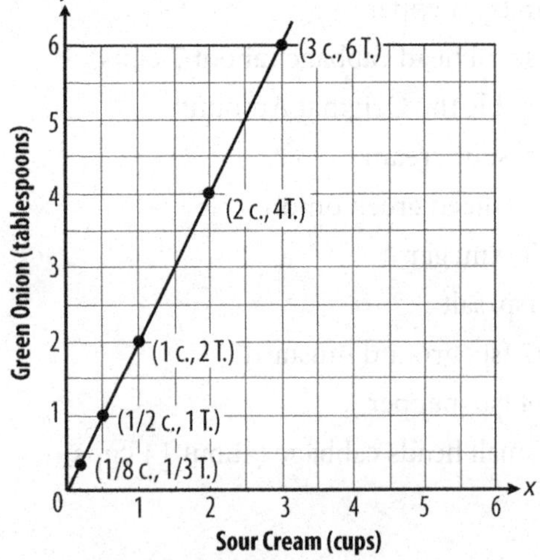

Task 3: How Low Can You Go?

1. 1 lb. Pasta: Foodtown (2/$1); $0.50/lb.
   Pasta sauce: Foodtown (2/$3); $1.50/jar
   Paper towels: All Foods ($5.49/6); $54.90
   Potatoes: Foodtown ($5.79/10 lb.)
   Ice cream: Foodtown ($1.89/qt.)
2. $64.58
   $0.50 + $1.50 + $54.90 + $5.79 + $1.89

Task 4: Test Practice

1. (4)
2. (4)
3. (1)
4. (1)
5. (4)
6. 16

Lesson 7: Comparing Walks

Activity 1: Team Walking Speeds

Answers will vary. See *Lesson 7 in Action* for examples of how teams made a table and responded to the questions.

Activity 2: Class Walking Speeds

1. Answers will vary.
2. Answers will vary.
3. Answers will vary.

Practice: Rounding Decimals from the Calculator Screen

Rounding to the nearest hundredth (0.01)

1. a. 0.333333333
   b. 0.33
2. a. 0.625
   b. 0.63
3. a. 0.809523809
   b. 0.81
4. a. 2.419354839
   b. 2.42
5. a. 0.062
   b. 0.06
6. a. 16.12903226
   b. 16.13

Rounding to the nearest tenth (0.1)

1. 0.3
2. 0.6
3. 0.8
4. 2.4
5. 0.1
6. 16.1

Rounding to the nearest whole number

1. 0
2. 1
3. 1
4. 2
5. 0
6. 16

Practice: Converting to the Unit Rate

1. 1 yd./sec.
2. 2.5 ft./sec.
3. 3.3 yd./sec. or 3 1/3 yd./sec.
4. 16 2/3 yd./sec. or 16.7 yd./sec.
5. 1.2 in./sec.
6. 0.7 ft./sec.
7. 0.75 ft./sec. or 3/4 ft./sec
8. 4. (16 2/3 yd./sec.)

Practice: Who Is the Fastest?

1. Joe was the fastest.

Joe	24 ft./8 sec.	3 ft./sec.
Seth	25 ft./10 sec.	2.5 ft./ sec.
Jorge	30 ft./14 sec.	2.1 ft./sec.

2. Tabitha was the fastest.

Shondra	5 mi./2 hr.	2.5 mi./hr.
Tabitha	3 mi./60 min.	3 mi./hr.
Narveen	3.5 mi./1 hr. 30 min., or	
	3.5 mi./1.5 hr.	2.3 mi./hr.
Ashlyn	5 mi./150 min., or	
	5 mi./2.5 hr.	2 mi./hr.

3. Morgan was the fastest.

Chee	6 laps/5 min.	1.2 laps/min.
Marsha	10 laps/8 min.	1.25 laps/min.
Morgan	8 laps/6 min.	1.3 laps/min.

Practice: MPH, MPG, and Other Rates

Estimates will vary. Some reasonable estimates are provided.

1. 450
2. 250
3. 10
4. 80
5. 75
6. 250
7. 330
8. 525

Extension: On the Treadmill

1. The incline and the speed
2. The number of calories, the distance, and the time
3. Yes. Answers will vary. Sample answer: The woman went twice as far and burned twice the calories. The distance/time or speed is 3.0 when you calculate 0.13 mi/2.45 min. = 0.05 mi./min. Because there are 60 minutes in an hour, that would mean 3 mi./hr. (60 x 0.05).

Calories	Distance	Time	Incline	Speed
About 40	About 0.5	10:00	0.0	3.0

Calories	Distance	Time	Incline	Speed
About 80	About 1.0	20:00 or 21:00	0.0	3.0

Calories	Distance	Time	Incline	Speed
250	About 3	About 60	0.0	3.0

Calories	Distance	Time	Incline	Speed
About 375	4.5	About 90	0.0	3.0

Extension: Graphing Treadmill Distances versus Times

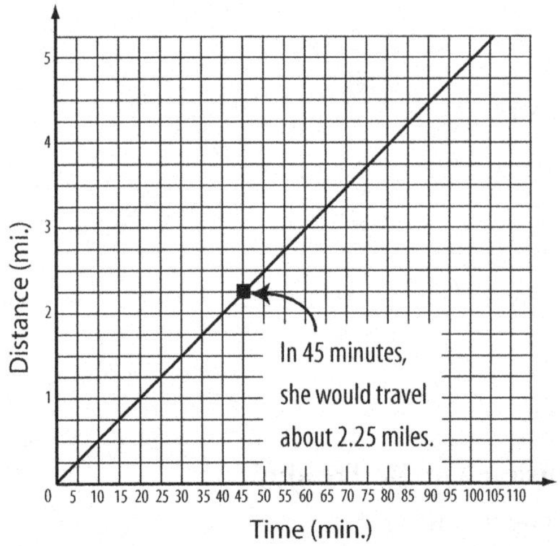

Test Practice

1. (3)
2. (2)
3. (5)
4. (5)
5. (4)
6. 240

Lesson 8: Playing with the Numbers

Activity 1: Creating and Checking Equal Ratios

1. **a.** Answers will vary. Sample answer: 50/100 = 30/60.

   b. Answers will vary. Sample answer based on 1a: 50 x 60 = 3000 and 100 x 30 = 3000.

   c. Answers will vary. Sample answer: 50 x 2 = 100 and 30 x 2 = 60.

2. **a.** Answers will vary. Sample answer: 50/100 = 60/30.

   b. Answers will vary. Sample answer based on 2a: 50 x 30 = 1,500 and 100 x 60 = 6,000.

   c. Answers will vary. Sample answer: 50 x 2 = 100 and 60 x 2 ≠ 30.

Activity 2: What Is the Fourth Number?

Card Set 1: Friendly Numbers

1. Answers will vary.
2. Answers will vary. Sample answer:

$$\frac{\boxed{20}}{\boxed{10}} = \frac{\boxed{5}}{\boxed{2.5}}$$

3. Reasons will vary. Sample answer: The proportion 20/10 = 5/2.5 is true. That can be verified by checking the relationship among the ratio numbers: 20 is double 10, and 5 is double 2.5; the relationship between the ratio numbers: 20 ÷ 4 = 5, and 10 ÷ 4 = 2.5; the unit rates are equal: 20 ÷ 10 = 2, and 5 ÷ 2.5 = 2; and the cross-product rule works: 20 x 2.5 = 50, and 10 x 5 = 50.

Card Set 2: More Challenging Numbers

1. Answers will vary.
2. Answers will vary. Sample answer: 251/76 = about 400/123.
3. Answers will vary. Sample answer based on example above: A good estimation could be 400 because 251 is more than 3 times 76, and 400 is more than 3 times 123. Also, 123 is less than double 76, and 400 is less than double 251.
4. About 406.
5. Answers will vary. Sample answers: Using the unit rate: 251 ÷ 76 = 3.3, so the unit rate is 3.3/1; 123 x 3.3 = 405.9. Using the property of equal fractions: 76 x 1.62 = 123, so 251 x 1.62 = 406.6. Using the cross-product property: 251 x 123 = 30,873, and 76 x 406 = 30,873.
6. Answers will vary.

Practice: Relating Multiplication and Division

1. If 35 x $\underline{4}$ = 140, then 140 ÷ 35 = $\underline{4}$
2. If 182 ÷ 13 = $\underline{14}$, then 13 x $\underline{14}$ = 182
3. If 15 x 15 = $\underline{225}$, then $\underline{225}$ ÷ 15 = 15
4. If 90 ÷ 12 = $\underline{7.5}$, then 12 x $\underline{7.5}$ = 90
5. If 16 x $\underline{6.5}$ = 104, then 104 ÷ $\underline{6.5}$ = 16
6. If $\underline{20}$ x 8.5 = 170, then 170 ÷ 8.5 = $\underline{20}$
7. If 400 x $\underline{0.5}$= 200, then 200 ÷ 400 = $\underline{0.5}$
8. If 750 ÷ 25 = $\underline{30}$, then 25 x $\underline{30}$ = 750
9. If 45 x $\underline{4.5}$ = 202.5, then 202.5 ÷ $\underline{4.5}$ = 45
10. Answers will vary. Sample answer: Multiplication and division are opposite operations.

Practice: What Is the Missing Number?

1. 120
2. 8
3. 60
4. 60
5. 150
6. 200

Practice: Estimate and Solve

1. Estimate: 50; More Exact Number: 53.2
2. Estimate: 300; More Exact Number: 313.2
3. Estimate: 120; More Exact Number: 119.8
4. Estimate: 40; More Exact Number: 39.9

Practice: Patterns for Predicting

Estimates will vary. Sample estimates provided.

1. Estimate: 650 ÷ 2 = 325 (close to 328), so 716 ÷ 2 = 358.
   Exact: 361.3
2. Estimate: 300 x 3 = 900 (close to 887), so 450 x 3 = 1,350.
   Exact: 1,337.9
3. Estimate: 180 ÷ 10 = 18 (close to 17), so 660 x 10 = 6,600.
   Exact: 7,221.2
4. Estimate: 960 ÷ 8 = 120 (close to 126), so 800 ÷ 8 = 100.
   Exact: 98.2

Practice: Story Problems

1. Estimate: $20 (281 is about half of 542, so $20 is about half of $40).
   Actual: $19.67
2. Estimate: 18 gallons (450 is about 1 1/2 times 330, so 18 is about 1 1/2 times 12).
   Actual: 17.3 gal.
3. Estimate: 15,600 feet, or about 3 miles (2,640 is about 1/2 mile).
   Actual: 3 miles
4. Estimate: $90 (60 is about 3/4 of 75, so 3/4 of $120 = $90).
   Actual: $93.00

Extension: Bringing Proportions to Life

1. Answers will vary. Sample answer: 251/76 = 406/123.
2. Answers will vary. Sample situation based on proportion above: If $251 were the cost of 76 notebooks, then 123 notebooks would cost about $406.

Test Practice

1. (3)
2. (5)
3. (3)
4. (3)
5. (2)
6. 25

Lesson 9: The Indian Ocean Earthquake and Tsunami

Activity 1: Tsunami Numbers in the News

1. A tsunami triggered by a very large earthquake off the coast of the Indonesian island of Sumatra on December 26, <u>2004</u>, has left more than 150,000 people dead and millions homeless. Countries hit hardest by the disaster include Sri Lanka, Indonesia, India, Thailand, and the Maldives. Almost 75% of the deaths occurred in Indonesia, estimated at <u>110,000</u>. Sri Lanka was second highest with about 20% of the estimated deaths, or <u>30,000</u> people lost that day. The rest of the deaths, approximately <u>10,000</u>, occurred in the other nine countries affected by the tsunami.

   The <u>20</u> foot wall of water, higher than a two-story building, swallowed entire villages. The tsunami waves were not only very high, they moved at a much faster speed than normal. These waves were comparable in size to those you see on some of the surfing movies; but those waves travel at 30 miles an hour, and the tsunami waves were moving more than fifteen times as fast at <u>500</u> miles an hour. The velocity of the force is what caused the destruction—a massive force that swept away everything in its path.

   Earthquakes are measured on a Richter scale that has a range from 0 to 12. This was a destructive earthquake measuring <u>9.0</u> on the Richter scale, the fourth worst earthquake in recorded history. A 6.0 on the scale is a pretty bad earthquake. The Richter scale does not use direct proportion, so an 8.0 is not twice as strong as a <u>4.0</u>; it is massively more. There are about <u>8,000</u> mini-quakes a day, which is almost three million per year.

2.

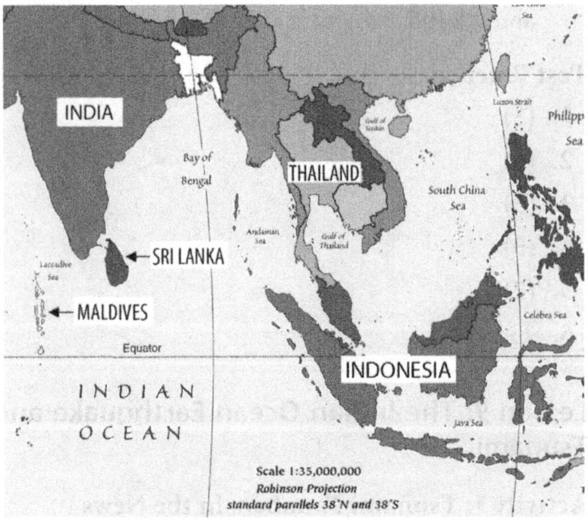

Activity 2: Tsunami Survivor Aid in Sri Lanka

1. For example, items that could be purchased for about 100 rupees each include an oil lamp, 10 candles, 10 boxes of matches, a pillow, a sleeping mat, a cooking pot, rice to feed a family of five for two days, half a pound of powdered milk, and two pounds of dried lentils. A small, three-room, cement-block home for a fisherman's family, Henning said, could be built for about 40,000 rupees; a fishing boat could be built for about 50,000 rupees; and an outboard motor could be purchased for about 100,000 rupees.
2. 5,000 rupees in donations is about $50. 10,000 rupees is a living wage. That is about $100.
3. Answers will vary. Sample answer: Multiply the number of dollars by 100.
4. Answers will vary. Sample answer: Divide the number of rupees by 100.
5. About $0.01. 100 rupees/$1 = 1 rupee/$0.01.

Activity 3: What Money Can Buy

Answers will vary. Sample answer provided:

1. Indonesia
2. Rupiahs
3. 9,298 (or 9,300) rupiahs = $1 US
4. Sample based on Indonesia:

   An emergency shelter kit ($225) = 2,092,050 (or 2,092,500).

   8 mosquito kits ($40) = 371,920 (or 372,000).

   A family food ration for 4 months ($200) = 1,859,600 (or 1,860,000).

   One month's supply of soap for 120 people ($20) = 185,960 (or186,000).

   70 packets of oral rehydration salts ($15) = 139,470 (or 139,500).

   Total ($500) = 4,649,000 (or 4,650,000) rupiahs.

Practice: Converting to U.S. Dollars

Answers will vary. Sample answers based on exchange rate on March 15, 2022.

1. 10 euros = $10.94
2. 100,000 yen = $845.20
3. 20 Canadian dollars = $15.66
4. 50 Mexican pesos = $2.40
5. 1 British pound = $1.30

Practice: The Cost of a Big Mac®

Answers will vary. Sample answers based on exchange rate on March 15, 2022.

1.

Country	Big Mac in U.S. Dollars
United States	$5.81
Great Britian	$4.68
China	$3.83
Indonesia	$2.37
Mexico	$3.31
Thailand	$3.82

2. Indonesia has the cheapest Big Mac; Great Britain has the most expensive.

Practice: At the Airport

1. $250 = about 190 euros (191.80)
2. 2.22 euros
3. Answers will vary. Sample answer: First find out about how many euros are in a U.S. dollar: 0.15 euro/$0.20 = 0.75 euro/$1.00. Then figure out how many yen are equal to $1 (which is the same as 0.75 euro): 1 euro/135 yen = 0.75 euro (or $1) = 101.25 yen. So there are about 100 yen in a dollar.

Extension: The Richter Scale and Earthquakes

1. The Indian Ocean tsunami earthquake was 10 times stronger than the San Francisco earthquake.

Extension: Your Own Country

Answers will vary, depending on students' experiences.

Test Practice

1. (5)
2. (2)
3. (3)
4. (5)
5. (2)
6. $42

Lesson 10: As If It Were 100

Activity 1: How Does Your Program Compare?

1. Answers will vary.
   Sample program data is provided.

Category	In Our Program	
	Number	% of total
Total Enrolled	275	100%
Gender		
Male	104	38%
Female	171	62%
Age		
16–18	15	5%
19–24	108	39%
25–44	96	35%
45–59	52	19%
60+	4	1–2%
Ethnicity		
Am. Indian/ Alaska Native	0	0%
Asian	21	8%
Native Hawaiian/PI	0	0%
Black /African Amer.	84	31%
Hispanic / Latino	70	25%
White	92	33%
More than one race	8	3%

Category	National Statistics	
	Number	% of total
Total Enrolled	1,100,210	100%
Gender		
Male	452,619	41%
Female	647,591	59%
Age		
16–18	92,028	8%
19–24	193,145	18%
25–44	538,578	49%
45–59	211,930	19%
60+	64,529	6%
Ethnicity		
Am. Indian/ Alaska Native	10,510	1%
Asian	109,630	10%
Native Hawaiian/PI	5,102	less than 1%
Black /African Amer.	193,653	18%
Hispanic / Latino	512,444	47%
White	245,273	22%
More than one race	23,598	2%

2. Answers will vary. Sample answer based on above sample program: In many ways our program is similar to the overall U.S. population, but there are also differences. Just as in the U.S. overall, our program is more female than male. However, we have relatively more women (62%) than in the overall U.S. population, where 59% of the population is female. As far as age groups go, our program has more people under 25 years old (44% in our program and 26% overall). However, in our program, the over-25 group tends to be a bit younger than in the U.S. Twenty percent of the people in our program are 45 or older, whereas 25% of the U.S. adult education population are over 45. The largest ethnic group in our

program is White (33%); yet in the U.S., it is Hispanic (47%). We have about the same percent of students identifying as more than one race (2–3%).

Sample circle graph comparison:

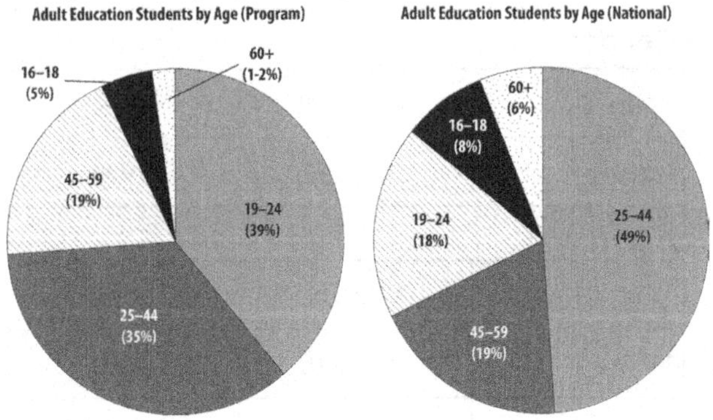

Activity 2: What If There Were 1,000?

1. Answers will vary.
2. Answers will vary. Sample report on languages: Because the 2022 world population is about 7,934,000,000, there would be about 14% (or about 1,110,760,000 people) who speak Mandarin. About 4% would speak Arabic, representing 317,360,000 of the world's population. People who speak Spanish would number 555,380,000 (or 7%), and 634,720,000 would speak English (or 8%).
3. Answers will vary.

Practice: Percent Estimates

1. 125 out of 250 = 50%.
2. 400 out of 1,200 = 33 1/3%.
3. 1,500 out of 2,000 = 75%.
4. 990 out of 1,000 = 99%.
5. 400 out of 1,200 = 33 1/3%.
6. 1,200 out of 2,400 = 50%.
7. 48 out of 480 = 10%.
8. 500 out of 2,500 = 20%.
9. 700 out of 7,000 = 10%.

Practice: Percent Dilemmas

Dilemma 1

Answers will vary. Sample explanation:
3 out of 6 = 1/2, which is the same as 50%.
50% means 50 out of 100, which is also 1/2.

Dilemma 2

Answers will vary. Sample explanation:
While there are 13,088 more multi-racial students than American Indian/Alaska Natives enrolled in adult education programs in the U.S., in percents, they both hover near 1% (American Indian/Alaska Natives = 1%; more than one race = 2%).

Dilemma 3

Answers will vary. Sample explanation:
If the ratio of teachers to students is 1 to 4, that means out of every 5 people, 1 is a teacher and 4 are students.
TSSSS = 4 out of 5 are students. 4/5 = 80%.

Practice: Free Throws—Who Is the Best?

1. Answers will vary. Sample answer if students looked at percentages:
   Tom made 64% of his free throws; Jerome, 45%; and Kayem, 75%. Therefore, Kayem did better than Tom, who did better than Kayem.
2. Answers will vary.

Practice: The Savings Contest

1. Answers will vary. Sample answer:
   Yes, Margot won the contest if you look at percent of each goal.
2. Answers will vary. Sample answer:
   If a student looks at this additively, Shaina is closest—only $300 away from her goal, as opposed to Tai, who is $440 away and Margot, who is $500 away. But Margot is 33 1/3 % away from her 100% goal; Shaina is 50% away, and Tai is 55% away. Based on percents, Margot is the closest to meeting her goal.

Practice: Pet Store Survey

Answers will vary. A sample report:
In a survey of 170 people, men seem to prefer dogs to cats 2 to 1; yet women prefer cats (62% are cat owners, but only 24% own dogs). Over half the women own cats, but only about 30% of the men own cats. Birds are much less popular with both men and women. Only 1 out of 10 men own birds, which is slightly less than women (at 13%). Younger adults and older adults differ. Cats are popular with those under age 35, where 54% own cats; but only 44% of those over 35 own cats.

Practice: Comparing Your Program to the Community Adult Education Program

Answers will vary. Sample answer based on the program sample for *Activity 1: In the Community Adult Education Program*, 42% of the students are male, while in our program, 38% are male. In both programs, about 31% of the adult student population is Black, or African American. Even though there are four students who are over 60 years old in each of the programs, this number represents about 5% of the Community Adult Education Program, but less than 2% of my program.

Extension: Parts per Million

1. Answers will vary.
2. **Good Products** / **Rejects**

	Good Products	Rejects
a.	99	1
b.	990	10
c.	9,900	100
d.	99,000	1,000
e.	990,000	10,000

3. Answer will vary. Sample answer:
   The number of good products increased and the number of rejects also increased. But the relationship remained the same: 99 to 1.
4. Answers will vary. Sample answer:
   No. 10,000 out of a million are too many to waste.

Test Practice

1. (5)
2. (3)
3. (5)
4. (2)
5. (2)
6. 69%

Closing the Unit: Reasoning with Ratios

Activity 1: Rates and Ratios in Your Life

Answers will vary. Sample response:
I investigated a gas station, where I bought gas at \$4.05 a gallon. I bought 12 gallons, so I paid \$48.60. This was a ratio that was proportional:

$$\frac{\$4.05}{1\text{ gal.}} = \frac{\$8.10}{2\text{ gal.}} = \frac{\$16.20}{4\text{ gal.}} = \frac{\$48.60}{12\text{ gal.}}$$

Activity 2: Final Assessment

Task 1: Statue of Liberty: The Nose Knows

Answers will vary, but should be close to the actual length of 42′. Students might reason from the picture or use the measurements of their own bodies. A reasonable estimate would be 40′–50′.

Task 2: Sampling M&M's

1. Answers will vary. Sample answer:
   There were 6 red M&M's out of a total of 58 M&M's in a bag that weighed 1.69 oz. In a 1-lb. bag, there would possibly be about 60 red M&M's, based on the sample from the small bag.

   6 M&M's/1.69 oz. = ? M&M's/16 oz. An estimate would be 60 because 16 oz. is about 10 times more than 1.69 oz. Using the cross-product property for a more exact answer, there would be about 57 red M&M's in a 1-lb. bag.
2. Answers will vary. Sample answer:
   No, there were 6 red M&M's and 13 brown M&M's. In a larger bag, there would be about twice as many brown M&M's as red ones, based on the sample from the small bag, or about 120.

Task 3: Making a Mixture

Roberto's mixture is stronger. Explanations will vary. Sample explanation:
Maia's mixture:

L/WWWWWWWWW

L/WWWWWWWWW

Roberto's mixture:

L/WWWWWWW

L/WWWWWWW

L/WWWWWW

In Roberto's mixture, there is less water per lemon concentrate (about 7 oz. water to 1 oz. lemon concentrate). Maia's mixture is 9 oz. water to 1 oz. lemon concentrate, so Roberto's mixture is more lemony.

Task 4: Traveling Buddies

Both cars will arrive at the same time. Explanations will vary. Sample explanation:
Bick's rate of 9 mi./12 min. is the same as 45 mi./60 min., which is the same as 45 mi. in one hour.

Task 5: Test Practice

1. (2)
2. (2)
3. (4)
4. (3)
5. (2)
6. 757

Sources and Resources

Mathematics Education

The National Council of Teachers of Mathematics (NCTM) publishes several excellent resources.

- *Principles and Standards for School Mathematics.* Reston, VA: NCTM, 2000.
- The NCTM journals: *Mathematics Teaching in the Middle Grades, Mathematics Teacher,* and *Journal for Research in Mathematics Education.*
- http://www.nctm.org

Fendel, D., D. Resek, L. Alper, & S. Fraser. *Interactive Mathematics Program.* Berkeley, CA: Key Curriculum Press, 1997.

Lappan, G., J. Fey, W. Fitzgerald, S. Friel, and E. Phillips. *Connected Mathematics Series.* Parsippany, NJ: Dale Seymour Publications, Division of Pearson Education, 1998.

National Research Council. *Adding It Up: Helping Children Learn Mathematics.* J. Kilpatrick, J. Swafford, and B. Findell (eds.). Mathematics Learning Study Committee, Center for Education, Division of Behavioral and Social Sciences and Education. Washington, DC: National Academy Press, 2001.

Russell, S.J., et al. *The Investigations in Number, Data, and Space Curriculum.* White Plains, NY: Dale Seymour Publications, 1998.

Mathematics and Numeracy Education for Adults

Curry, D., M.J. Schmitt, and S. Waldron. *A Framework for Adult Numeracy Standards: The Mathematical Skills and Abilities Adults Need to Be Equipped for the Future.* Boston: World Education, 1996. (This framework was developed by members of the Adult Numeracy Network.) https://adultnumeracynetwork.org/

Massachusetts Department of Education. *Massachusetts Adult Basic Education Curriculum Frameworks for Mathematics and Numeracy.* Malden, MA: Massachusetts Department of Education, 2001. http://www.doe.mass.edu/acls/frameworks

Stein, S. *Equipped for the Future Content Standards: What Adults Need to Know and Be Able to Do in the Twenty-First Century.* ED Pubs document EX0099P. Washington, DC: National Institute for Literacy, 2000.

Proportional Reasoning

Billings, E. "Problems That Encourage Proportion Sense." *Mathematics Teaching in the Middle School 7*, no. 1 (2001):10–14.

Billings, E. "Cocoa." In G. Bright and B. Litwiller (eds.). *2002 NCTM Yearbook: Classroom Activities for Making Sense of Fractions, Ratios, and Proportions.* Reston, VA: NCTM, 2002, 38–40.

Clement, L. "Make a New Puzzle." In *Classroom Activities for Making Sense of Fractions, Ratios, and Proportions.* G. Bright and B. Litwiller (eds.). Reston, VA: NCTM, 2002, 43–45.

Cramer, K. and T. Post. "Making Connections: A Case for Proportionality." *Arithmetic Teacher,* 40 (February, 1993):342–46.

Lamon, S. J. "Ratio and Proportion: Children's Cognitive and Metacognitive Processes." In *Rational Numbers: An Integration of Research.* T. Carpenter, E. Fennema, and T. Romberg (eds.). Hillsdale, NJ: Lawrence Erlbaum Associates, 1993, 131–196.

Lamon, S. J. "Part-Whole Comparisons with Unitizing." In *2002 NCTM Yearbook: Making Sense of Fractions, Ratios, and Proportions.* B. Litiller (ed.). Reston, VA: NCTM, 2002, 79-86.

Langrall, C. W. and J. Swafford. "Three Balloons for Two Dollars: Developing Proportional Reasoning." *Mathematics Teaching in the Middle School 6*, no. 4 (December, 2000): 254–6.

Lave, J. *Cognition in Practice.* Cambridge, UK: Cambridge University Press, 1988.

Lo, J. and T. Watanabe. "Developing Ratio and Proportion Schemes: A Story of a Fifth Grader." *Journal for Research in Mathematics Education 28*, no. 2 (1997): 216–23.

Lanius, C. and Williams, S. "Proportionality: A Unifying Theme for the Middle Grades." *Mathematics Teaching in the Middle School 8*, no. 8 (2003): 392–396.

Smith, J. P. "The Development of Students' Knowledge of Fractions and Ratios." In *2002 NCTM Yearbook: Making Sense of Fractions, Ratios, and Proportions.* B. Litiller (ed.). Reston, VA: NCTM, 2002: 3–14.

Weinberg, S. (2002). "Proportional reasoning: One problem, many solutions." In *2002 NCTM Yearbook: Making Sense of Fractions, Ratios, and Proportions.* B. Litiller (ed.). Reston, VA: NCTM, 2002: 138–144.

Context-Specific Sources and Web Sites

Schor, J. The Overspent American: Why We Want What We Don't Need. New York: Basic Books, 1998.

For international currency conversions, see

http://www.oanda.com/convert/classic

For information on the 2004 Indian Ocean tsunami, see

https://www.unicefusa.org/mission/emergencies/tsunamis/2004-south-asia

https://reliefweb.int/

https://www.oxfamamerica.org/

Made in the USA
Monee, IL
08 November 2026

68032037R00129